MOTORBOOKS POWERTECH SERIES™

HOW TO TUNE & MODIFY CHEVROLET TPI ENGINES

T. + J.

2003

JASON SCOTT

Dedication

For Goodwin Hannaford and Frank Juliano.
Thank you both for all you've taught me about engines, cars, and the people who enjoy them.

First published in 1999 by MBI Publishing Company, 729 Prospect Avenue, PO Box 1, Osceola, WI 54020-0001 USA

MBI Publishing Company books are also available at discounts in bulk quantity for industrial or sales-promotional use. For details write to Special Sales Manager at Motorbooks International Wholesalers & Distributors, 729 Prospect Avenue, PO Box 1, Osceola, WI 54020-0001 USA.

Library of Congress Cataloging-in-Publication Data

Scott, Jason.
How to build & modify Chevrolet TPI engines / Jason Scott.
p. cm.
Includes index.
ISBN 0-7603-0573-0 (pbk. : alk. paper)
1. Chevrolet automobile—Motors—Modification. 2. Chevrolet automobile—Motors—Fuel injection systems. I. Title. II. Series.
TL215.C5S44 1999
629.25'04—dc21 99-19843

On the front cover: The Chevrolet TPI engine is used in the bowtie brand's most powerful cars. The Corvette benefits from Tuned Port Injection to drive its V-8 powerplant.

Designed by Bruce Leckie

Printed in the United States of America

CONTENTS

ACKNOWLEDGMENTS

Writing a book is a lot like building an engine. You work with a lot of people getting parts and advice, and even joining you in marathon wrench-bending sessions. This book is no different, and may actually be even more similar, since it deals with building engines.

In any event, there are a lot of people who directly or indirectly contributed to making this book what is.

The top people on my list for thank yous are my daughter, Jenna, and my wife, Michelle. Through their gracious gifts of time and sometimes solitude, I was able to boil down my copious notes into what I hope are clear, comprehensible discussions of the various topics covered.

Though I consider myself to be pretty well versed in engine engineering theory, I am by no means an expert on the subject. I still get confused. I still make mistakes. I rely on other people to set me straight so that I don't send you off on a wrong path, wasting both your time and your money.

Some of the most helpful people on this project were the kind folks at Callaway Cars. They allowed me virtually free roam of their Old Lyme, Connecticut, facilities. For that, I'm eternally grateful. In between my moments of feeling like a kid in a candy store, surrounded by exotic crankshafts, connecting rods, and cylinder heads, the Callaway engineering wizards shared some of their experience and knowledge with me. For all that, I extend my sincerest thanks to Herb Couzins, Thomas Weiss, Mike Zoner, Bill Prout, and, of course, Reeves Callaway. I simply couldn't have completed this book without their assistance.

I am equally indebted to Myron Cottrell of TPI Specialties and Cottrell Racing Engines. Myron's practical, no-nonsense advice appears throughout this book. He pointed out some common-sense approaches to typical performance build-ups and generously helped me understand some of the finer points of race engine mechanics.

The flip side of that is the production end of engine development, and for that I've relied on an extensive list of GM's powertrain and vehicle engineers, including Jim Minniker, John Heinricy, Rod Michaelson, Jim Ingles, Jon Moss, and others. Without folks like these, we wouldn't have TPI engines to flog and enjoy.

And though my conversations with them weren't always part of my research for this book, both John Lingenfelter and Doug Rippie laid the groundwork of my understanding of TPI engines and ultimately influenced every page of this book. The times I pestered them at various events and over the phone were some of the most thought-provoking and helpful times I could have hoped for.

There are literally dozens more people who contributed in large and small ways to this book, and to all of them I extend my thanks. But it would be a particularly gross oversight on my part to ignore the many fine folks at MBI Publishing, specifically Zack Miller and Chris Batio. They've given me the opportunity to spend time talking cars and engines with some of the best people in the country—and to pass it off as work.

Thanks everyone!

INTRODUCTION
DON'T GRAB YOUR WRENCHES QUITE YET

When someone asks me what they should do to their TPI engine to increase its power, my standard response is, "What do you want to do a foolish thing like that for?" After all, even the weakest TPI engine is more than powerful enough to move a car faster than any speed limit in the United States will allow.

More than that, General Motors' powertrain engineers are pretty smart folks and they have done a fine job of delivering not only power but efficiency, durability, and reliability too—a complete, well thought-out performance envelope. Sure, tampering with the configurations of the production engines can produce gains in one area or another, but it's going to be awfully tough to achieve the sort of balance the factory has achieved. Any gains you make are bound to be at the expense of one or more of the other aspects of performance.

I'm not usually too quick to tell people what to do to increase the power of their TPI engine. That's why the rest of this book is basically against my better judgment. But if you simply must modify your engine (and obviously you feel you must, or you wouldn't have bothered picking up this book) there are a few things you should know first.

Do We Understand Each Other?

A few words have to be said about the different TPI engines and terminology mentioned throughout this book. Strictly speaking, the only engine Chevrolet ever called "tuned port injection" (TPI for short) was the 1985–1992 305/350 with the long, curved runners. The ZR-1's LT5 engine, the LT1/L99/LT4 family, and the new LS1 and Vortec truck engines are not technically TPI engines, but since they are all so similar in design and share many parts, the term TPI is used to identify one or all of them throughout this book.

Some additional confusion can arise because of General Motors' propensity to develop different generations of its engines. Technically speaking, the first-generation small-block Chevy wasn't called anything except a small-block Chevy V-8 during its entire life, from 1955 through today. And with the exception of the change to a one-piece rear main seal in 1987, and the switch to roller cams in 1988, a 1955 engine block is pretty much the same as a 1998 block. Throughout this book you will usually find these engines referred to as first-generation engines, or sometimes Gen-I.

In 1992, GM introduced the LT1, which was the first member of the Gen-II (for second-generation) engine family. Though still very similar in design to the first-generation engines, the Gen-II featured enough changes that few parts are interchangeable. The block, crank, cylinder heads, intake manifolds, and other parts are specific to the Gen-II family.

Making things slightly more confusing is that GM also released an engine it dubbed the "SB2," for small-block, version 2. SB2 and Gen-II engines have almost nothing in common. Furthermore, the SB2 is primarily intended for racing use, such as in NASCAR's Winston Cup ranks. Its primary departure from previous small-blocks were the symmetric-port cylinder heads, which interestingly enough, turned up in the next small-block iteration.

An all-new V-8 appeared beneath the curvaceous hood of the then all-new 1997 Corvette. This engine—and the

A well-planned and built high-performance TPI engine, like this Callaway Cars Super Natural 383-cubic inch LT1, will produce gobs of power, be rock-solid reliable, exceptionally efficient, and pretty darned good looking too. Callaway's SuperNatural LT1 makes 404 horsepower on pump gas.

1999 Vortec 4800, 5300, and 6000 V-8s based on it—is loosely referred to as "Gen-III" or "third-generation" small-block Chevy V-8s, though they share nothing more than a few fasteners in common with previous engines. The highlights of the Gen-III motors are their deep-skirt blocks with main bearing caps that are retained by six bolts, plus the symmetrical-port cylinder heads with 15-degree valves. Intake manifolds for Gen-III engines are of a molded composite material. Other upgrades abound throughout the engine.

Keeping these major eras of the small-block in mind as you read through this book will help you sort out what modifications can and should be made to what engines. But bear in mind that each era has its own subtle variations, so parts don't always interchange as you might expect—even ones that look quite similar.

Actual Mileage May Vary

There are many products mentioned in this book, and there are even more engine engineering concepts discussed. Each has been proven—if not by me, by one of the many people or companies with whom I collaborated to produce the book—to be effective under certain conditions.

Some parts only work well with certain other parts, or in certain specific situations. Likewise, some modifications are better suited to some applications, driving conditions, or types of components. For example, opening up small-port cylinder heads to increase airflow generally yields measurable performance gains, but opening up large port heads to flow even more air than they already can may actually decrease power because you can make the ports too big for your engine's configuration.

The key to all this is to choose the right combination of parts. Some companies actually make this easy for you, by doing all the engineering and testing work in advance, so you can purchase a complete solution: cam, heads, and intake that will all work harmoniously. While such a kit may not seem as exciting as picking out a custom assortment of parts from different manufacturers, you're assured the kit's components will all function well together. It takes all the guesswork out of things, and it's a method I highly advocate—especially for beginners. You'll know exactly how much power your engine should be making when you're done, and if you encounter problems, you'll have one phone number to call for tech support and shouldn't get the runaround of one manufacturer blaming another for a given problem.

If you opt to go the piecemeal route, consider yourself warned that your actual mileage—and performance, reliability, and durability—may vary.

Some Assembly Required

Ultimately, this book wasn't designed to spell out every change you should make to your engine. Instead, it is intended to help you better understand your needs, and thus those of your engine, and how to address those needs with whatever performance parts and modifications may be available to you at the time you set about making changes to your engine.

With any luck, you will be educated well enough to understand the impact new parts and technologies will have on engine performance as they become available. The automotive industry moves pretty fast, and who wants to be bolting on yesterday's recommended parts when something newer and better may be available?

So, think your proposed changes through carefully, because the decisions and changes you make are your decisions—they're your choices. Enjoy the experience. After all, half the fun is in making the changes.

— *Jason Scott*

CHAPTER ONE

TPI'S PAST, PRESENT, AND FUTURE

It has been said that history repeats itself. That may be true, but in the case of modifying Chevrolet TPI or other fuel-injected Chevy engines, that's not always the most efficient or productive exercise. Taking a few moments to familiarize yourself with the TPI engine's history can provide a few clues as to why things are the way they are, so you don't haphazardly change things that would be better left alone.

Likewise, it's rarely a good idea to go off half-cocked on an engine build-up. Failure to define the goals for your engine makes it almost impossible to map out a plan that will ensure you attain every one of them.

In a nutshell, the sole purpose of this chapter is to help you with both of those issues. Our quick look back at Chevy's various fuel-injected engines should give you the background you need to know just what the heck it is that you're about to dig into. You'll also likely glean some insight into what sorts of setups have been tried, which have worked, which haven't, and therefore what types of modifications you might want to consider.

Planning-wise, things boil down to knowing what you want, then methodically setting out on a course that will help you get it—whatever "it" may be. In general, this means figuring out how the engine will be used, what duties it will be called upon to perform, and what sorts of punishments it will need to endure.

In the end, the few minutes you spend now could save you countless hours and hundreds—if not thousands—of dollars later, correcting for mistakes and poorly engineered modifications.

Chevy's EFI Gene Pool

The roots of Chevrolet's electronically controlled fuel-injection engines run

Chevrolet's original entry into the world of fuel injection was the 1957 Rochester Ramjet system for the 283-cubic inch small-block Chevrolet V-8. With fuel injection, the 283 produced 283 horsepower. Ironically, many buyers wound up swapping the "Fuelie" setup for a traditional carburetor for easier maintenance and tuning. The Ramjet system stuck around on Corvette's until 1965.

It wasn't until 1982 when Chevy took another whack at fuel injection, with the ill-named "Cross-Fire Injection" electronically controlled throttle-body injection system on Corvettes and the redesigned Camaro Z28. The Cross-Fire system also metered and mixed fuel for the redesigned 1984 Corvette's 350.

deep. Chevy's first foray into the world of fuel injection appeared atop the 1957 Chevrolet small-block in the form of the Rochester-designed RamJet system. By today's standards, it was a crude system—though incredibly complex—and, in truth, it really wasn't much better than a carburetor. In fact, many enthusiasts would argue it was a lot worse.

The RamJet system stuck around through the 1965 model year on Corvettes, then disappeared from option lists forever.

During much of the 1970s, Chevy—and every other automobile manufacturer—had its hands full trying to make its engines meet new emissions regulations, among other mission-critical projects. As a result, it didn't have the engineering brainpower to expend on developing radically different new induction systems; nearly every engineer on the payroll (and then some) were frantically struggling to maximize the efficiency of the existing carburetion, ignition, and exhaust systems technology. Computer technology was still in its infancy and cost-prohibitive, as well. For the most part, engineers had to make do with what they had; pioneering new technologies would have to wait until the company had a better grasp on what was needed to make the engines more efficient, and there was no other way to achieve cleaner emissions without new technology.

Toward the late-1970s, GM saw the handwriting on the wall for its re-engineered carburetion and ignition systems; they wouldn't be able to meet the more stringent emissions regulations that were coming. So GM stepped up development of what turned out to be little more than glorified electronic carburetors. Its 1982 return to the world of fuel injection was given the dubious name: Cross-Fire Injection, which many believed sounded more like a malady than a cure. The system fitted two twin-venturi throttle body units to a cross-ram-style aluminum intake manifold. Each throttle body unit was similar to a carburetor, except that fuel didn't run through it; instead, an electronically controlled fuel injector sat perched atop each venturi and was fired by a computer, based upon input from a number of sensors.

Cross-Fire Injection looked and sounded impressive, and it worked. But the system was complicated, and it still suffered from many of the same problems as carbureted engines—uneven air/fuel delivery to each individual cylinder, relatively poor fuel atomization, fuel puddling in the intake, and others. Clearly, a better solution was still needed.

In 1985, Chevrolet introduced the Tuned Port Injection system in the Corvette. It wasn't just revolutionary automotive technology, it was sexy. It looked radically different from either of the company's earlier fuel-injection attempts, thanks to a long, finned plenum with large, long, curvaceous runners. And the rest of the engine looked equally exciting. Gone was the maze of vacuum lines, check valves, and distribution tees. The TPI engine showed that GM had learned a thing or two about engine appearance.

Whereas all earlier fuel-injection systems were "wet" systems—meaning both air and fuel traveled through the intake manifold—TPI systems flowed only air through the intake. Fuel got added by eight individual fuel injectors—one mounted in the end of each intake runner, positioned to spray toward the intake valve. This fundamental difference means that you don't have to worry (as much) about fuel atomization, fuel won't ever puddle up in the intake, and air/fuel mixtures can be optimized for each cylinder. Relieved of restrictions those problems impose on intake manifold design, engineers were free to shape the intake runners to maximize air delivery, knowing the injectors and computer controls would optimize the air/fuel ratio, thus maximizing power.

Since street engines benefit most from increased low- to mid-rpm torque, the engineers gave the TPI system long runners, roughly 18 inches each. And since fuel was introduced just ahead of the intake valves, it was easy to ensure that each cylinder got the same air/fuel mixture, eliminating the age-old fuel-distribution problem that resulted in cylinders nearer to the carburetor or throttle body injection system being richer than more distant cylinders. This change decreased fuel consumption and emissions output while simultaneously increasing power output.

While the TPI engines were clearly showing the way of the future, GM continued to refine the less expensive throttle body injection (TBI) systems. Though essentially still similar to a carburetor on a standard intake manifold, TBI did provide more precise fuel metering and an improved spray pattern for better fuel atomization, which aided both fuel efficiency and power. And naturally, intake manifold designs were optimized for the

Chevy made a quantum leap from the "glorified carburetor" Cross-Fire Injection system to the all-new multipoint-injected 1985 L98 Tuned Port Injection 350 for the Corvette and the similar LB9 305-inch version for the Camaro. The engine looked radically different, with its boxlike plenum and long, shapely runners. A fuel injector in each intake runner mixed fuel with the intake air just prior to its entry to the cylinder head, ensuring an optimal fuel mixture for each cylinder.

The TPI engine was a marvel of engineering, but it was also expensive. Chevy opted to stick with throttle-body injection for its truck line, which still resulted in a significant improvement in fuel efficiency, reliability, and emissions quality.

improved TBI systems. TBI systems were flexible and easy to adapt to different engines too. GM fit TBI systems to everything from four-cylinder compact economy cars to massive 7.4-liter big-block V-8-powered trucks.

The next major overhaul of the TPI system came in 1992, when the second generation Chevrolet small-block V-8 debuted: the LT1. With 300 net horsepower, the LT1 was a powerhouse that actually made more power than the legendary old-world small-block for which it was named: the 1970 LT-1.

The LT1's most obvious difference was its intake manifold design. Instead of the TPI's long runners, the LT1 had short, stubby runners that were only a little more than 3 inches long. Just like a single-plane carbureted intake manifold, these short runners boosted breathing at higher rpm, and because of camshaft and ignition advances, the engine gave up little in terms of low-rpm power.

But there were far more changes with the LT1, which truly justified the Gen-II (generation two) moniker it was given. One of the unique changes was the incorporation of a reverse-flow cooling system that routes coolant to the cylinder heads first, then into the block—the opposite of previous small-block engines. This change, plus redesigned combustion chambers, allowed the use of higher compression—10.25:1—which further contributed to power.

Also, the new Opti-Spark ignition system replaced the traditional distributor with a crank-triggered system that was accurate to within 1 degree of crank rotation, which made it far more precise than a mechanical distributor could ever have hoped to be.

As good as the Gen-II engine family was—the last hurrah, the LT4, pumped out 340 horsepower in stock trim—there were numerous limitations of the engine's basic design. It was too heavy, too tall, had some less-than-ideal valvetrain geometry, and other constraints that just couldn't be overcome in an evolutionary fashion. A clean-sheet design was going to be the only solution.

Meanwhile, the truck line engineers weren't yet ready to give up on the venerable old small-block. They redesigned cylinder heads with better ports and chambers, revised camshaft profiles, updated the induction system to a more TPI-like system with a central air-metering throttle body and individual poppet-valve injectors, plus a unique two-piece intake manifold that featured a composite top half mated to an aluminum base. A compression bump and new cam profiles raised power output and broadened the torque curve but reduced fuel consumption and emissions, making the engines successful from any perspective.

But these Vortec V-8s, as successful as they were, still only delayed the

TPI to the max: the ZR-1 Corvette's 5.7-liter LT5 V-8. The all-aluminum dual overhead cam engine was designed and built in conjunction with Lotus and featured a number of trick features, including secondary intake runners for each cylinder that opened as needed to ensure significant high-rpm power: 375 horses in 1990 when it officially debuted, and 425 by the time it disappeared from the option sheets in 1994.

In 1992, the second-generation—or Gen-II, for short—small-block Chevy V-8 debuted under the Corvette's clamshell hood. Gone were the first TPI system's long, curving runners. In their place was an extremely short runner intake, plus redesigned cylinder head ports, a front-mounted ignition distributor, a reverse-flow cooling system, and other improvements that made for 300 horsepower from the familiar 350-cubic inch displacement.

inevitable; the aging small-block was reaching the end of its useful life. A replacement small-block was already well underway.

The engine that ultimately resulted is known as the LS1 engine, and it made its first appearance in the 1997 Corvette, followed shortly by use in the Camaro and Firebird performance models. About the only thing the LS1 shares with either of the Gen I/II engine families is the almost sacred 4.40-inch bore spacing that small-block Chevys have had since 1955. Everything else is all-new, from its aluminum oil pan to its composite (meaning, plastic) intake manifold, and everything in between, including the gaskets. In fact, the only parts shared between an LT1 and an LS1 are a few miscellaneous fasteners.

A new crankshaft was designed with revised everything—journal sizes, stroke length, oiling passages, you name it. Connecting rods continued to be produced from powdered metal—a process GM pioneered on the LT1—but were of a totally new design, with all-new dimensions. Pistons, likewise, were new. The new aluminum cylinder heads featured symmetrical, high-flow ports, powdered metal roller rocker arms, and a radically different intake valve angle, compared with production Gen-I and Gen-II small-block heads.

Nestled between the heads was a new, injection-molded plastic intake manifold with medium-length, cross-flow runners and a new single-blade throttle body assembly. Meanwhile, a new ignition system was triggered off both a crankshaft position sensor and a separate camshaft position sensor to fire each of the eight ignition coils—one for each spark plug.

The new engine addresses nearly all of the old engines' shortcomings. It is lighter, smaller, and more fuel efficient, yet produces more power per liter. It has a broader powerband; a higher redline engine speed; a better oiling system; a better ignition system; fewer coolant, oil, and air leaks thanks to the new reusable gaskets; cleaner emissions; increased reliability; and the list goes on and on. With the introduction of the 1999 Chevy Silverado and GMC Sierra pickups, there is even a trio of cast-iron Gen-III small-blocks: the Vortec 4800, Vortec 5300, and Vortec 6000, which replace the 5000, 5700LD, and 5700HD engines, respectively.

At the time this is being written, about the only area in which the LS1 isn't as attractive as the earlier engines is in the aftermarket parts market, where relatively few performance parts are available. But that's sure to change as the world realizes and develops an appreciation for the many advantages the LS1 offers to the performance world.

Mapping Your Route

As the saying goes, "There's more than one way to skin a cat." So, too, there's more than one way to build a high-performance Chevy TPI engine. Depending on your outlook on life, there can be an infinite number of approaches, or as few as just two: the Right Way and the Wrong Way.

Professional engine builders tend to agree that you can't put together a winning engine if you don't know what kind of competition it will be facing. As an extreme example, an engine set up for powering the truck you haul your car trailer, boat, or camper with won't stand a chance in a road race. Likewise, an engine built to run 10-second quarter-miles isn't going to serve you well as a daily driver, nor will it likely last long under such conditions. The name of the game is to figure out how your engine will be used, then build it to meet—and exceed—the demands of that application.

While there are doubtless other possibilities, for our purposes there are essentially four distinct engine applications for which you might custom-build a TPI engine: high-performance street, drag racing, autocrossing, and road racing. Of course, there are also several degrees of modification for each application.

Each of these applications has its own unique requirements and restrictions, be they the need to withstand high rpm, or simply rules that prevent certain modifications. If you take the time to understand what the demands will be on

The Vortec V-6 was Chevy's first multipoint fuel-injection system for its truck line and originally featured a familiar-looking plenum and runner system.

your engine, and any restrictions or limitations you are confronted with, you can select the appropriate parts and perform the modifications that make the most sense.

Following are brief summaries of the typical needs of the basic applications, as well as general recommendations of what your engine should be set up to handle for each, and even a few guidelines of the types of modifications to make.

Chevrolet later redesigned the Vortec engines to use a less-expensive but just-as-effective central fuel-injection system, with an air-metering throttle body atop a unique two-piece plastic and aluminum intake manifold. A single fuel injector distributes fuel to a poppet valve mounted in each intake runner for precise fuel mixing. The new Vortec system was grafted onto all gas-powered truck engines from the 4.3-liter V-6 to the mighty 7.4-liter (454-cubic inch) big-block. For direct comparison, the 350 produced 255 horsepower and 330 pound-feet of torque in 1998.

Street High Performance

Bowing to the gods who proclaim that Bigger Is Better, there's no better setup for the street than a stroker 377- or 383-inch small-block, whether carbureted or fed by an EFI system. The rewards of EFI, of course, are generally more torque, increased efficiency, improved reliability, and reduced maintenance, though carbs can often surpass at least the stock TPI systems in terms of peak horsepower. Naturally, you should strive to keep your vehicle emissions legal, which will likely mandate EFI, just to minimize emissions output.

Barring a stroker package, the production 350 makes the most sense, since the 305 simply isn't designed for true high-performance use, and there's little point in starting yourself out with a disadvantage, in terms of torque and horsepower potential.

So where does one begin? Fortunately, the stock bottom end should be good for 400 or so horsepower, according to Myron Cottrell of TPI Specialties, assuming, of course, that it is in good condition. Thus, leaving the bottom end alone, Cottrell recommends swapping the cam first, to go for the big gain right off the bat, rather than messing around with little gains here and there, as you would get with new air filters, plug wires, and other bolt-on upgrades.

While you've got the engine torn apart, it's a good time to swap the heads (what's another couple of dozen bolts per side at that point?) for either CNC-machined parts or custom-ported units, or if you're on a budget, pull yours and have them reworked with a multi-angle valve job and new springs and port-match them to your intake. For the street, valve sizes should be kept relatively small: 1.94 intakes and 1.60 exhausts will generally provide better breathing, without being so big that they end up killing low-end torque. Headers are easier to install with everything apart, too, so go for a set of ceramic-coated pipes, preferably of equal length. And you might as well swap the intake base for a better flowing unit before you get everything bolted back together.

As the intake is going back together, install an adjustable fuel pressure regulator and set it to deliver 45 pounds or so, for a baseline, and adjust up or down depending on what driving results you get.

Next, you should concentrate on helping the engine breathe better. A cold-air induction system with a free-flowing air filter will help feed the engine, while a cat-back exhaust system—one that replaces the pipes and mufflers from the catalytic converter's outlet rearward—will keep it from getting plugged up. If you happen to have a carbureted engine, give some thought to switching to an EFI system. A simple throttle-body setup like Holley's Pro Jection system delivers excellent performance, or ACCEL has several kits available that offer even more advanced features.

From there changes become less dramatic. A high-quality ignition system will make sure as much fuel as possible gets burned. Reprogramming the vehicle's computer, known as either an ECM (Engine Control Module), PCM (Powertrain Control Module), or VCM (Vehicle Control Module), depending on the year and what vehicle systems with which it interacts, will help you optimize fuel delivery and ignition timing, among other parameters, to improve engine efficiency with the modifications. ECMs store engine operating parameters in a PROM (Programmable Read Only Memory) chip, which can be easily replaced with a version with different, high-performance parameters, while PCMs and VCMs can simply have new instructions uploaded to their EEPROM (Electronically Erasable Programmable Read Only Memory) chip set.

Again, whatever modifications you choose to make, you should strive to be conservative with them, choosing to err on the side of caution. A cam that's too big will kill performance quickly, while a cam that's slightly too small will still generally provide excellent performance.

Drag Racing

Drag racing requires an engine that can accelerate quickly, has lots of torque for quick launches, as well as plenty of horsepower for good top speed through the traps. Again, a stroker motor makes an excellent choice, but the production 350 is an outstanding starting point and may be your only viable choice, depending on class rules.

For serious drag racing, you can't go wrong rebuilding the bottom end with forged pistons, zero-gap rings, a steel crank, heavy-duty connecting rods, and a heavy-duty oil system, all in the name of increased durability to face the tremendous internal stresses the engine will encounter.

A cam swap and freer-breathing cylinder heads make excellent additions, too, while a set of long-tube, equal-length headers will yield the most power. In terms of induction systems, you want something that will flow well at high rpm (above 5,000 rpm), such as a short-runner intake like the LT1/LT4 unit or the remarkably similar TPI Specialties Mini-Ram intake (which works on first-generation small-blocks that require a cooling system crossover). TBI engines should respond well to a tall-profile dual-plane intake, or preferably a single-plane intake.

Drag-racing engines place high demands on ignition systems, so don't skimp here. A high-output coil, multiple-discharge spark box, and low-resistance wires should be a minimum, and you'll probably want a multistage, rev-limited timing computer, and possibly a retard device if you'll be running nitrous.

Nitrous is almost a given for a serious drag-racing engine. The output level you choose depends upon bottom-end modifications, but the most efficient nitrous systems have direct port injection, plus their own fuel delivery system, including a dedicated fuel pump.

The stock cooling system should generally be up to the task of cooling your drag-racing engine, though addition of an electric water pump drive kit will allow the cooling system to keep doing its job after you shut the car off back in the pits.

Autocross

Ultimate engine output isn't as crucial in autocross racing, so preparations for it should be similar to those for street racing: a good cam, some head work or replacement heads, some intake porting and port-matching or a replacement intake, headers, ignition, and so on. The name of the game should be maximizing usable torque and improving throttle response. Steer clear of nitrous, superchargers, turbos, and other such toys; they'll just cause the tires to go up in smoke and increase your lap times.

Road Race

Road racing will place the highest demands on an engine because it must be docile enough to come down to low rpm for tight corners, then accelerate back up to high rpm over and over, lap after lap

The focus, here, needs to be on durability. If you choose to rebuild the bottom end (and it's a wise idea), install forged aluminum pistons with high-quality rings; have the rods rebuilt or—better yet—replace them with higher strength aftermarket parts and top-quality rod bolts; drop in a steel crank that has been cross-drilled for better oil flow; and run four-bolt main caps held in place with studs (on the center bolts).

The camshaft will need to meet class rules, but several are available that qualify as stock replacements. Install the cam with a high-quality timing set (a belt drive will allow a great deal of tuning flexibility, but a double-row true roller chain and gear set will work fine). Oversize 7/16-inch pushrods will improve top-end durability, while roller rockers will ensure accurate, nearly friction-free valve actuation. Racing spring retainers and keepers will keep the valves in place in the heads, which should be ported and massaged to the fullest extent of the rules and secured by studs. Headers will be necessary (rules permitting) for evacuating the exhaust gases, as will a free-flowing exhaust system.

A cold-air/ram-air induction kit and free-flowing air filter should feed an oversize throttle body connected to a ported intake.

The ignition system will also need to be upgraded with a high-output coil, multiple-discharge spark box, and a rev-limiter to keep you from getting too overzealous.

Your oil system should incorporate a high-volume oil pump, a windage tray, an extra-capacity oil pan, an oil cooler, and synthetic oil, while the cooling system should have an aluminum racing radiator and high-flow water pump as minimum equipment.

The third-generation (aka, Gen-III) small-block Chevy engine arrived under the hood of the redesigned 1997 "C5" Corvette. Still displacing 5.7 liters (actually it's 5.65, but who's counting?) the Corvette's all-aluminum LS1 churned out 340 horses and features a number of significant improvements compared to the small-blocks of old, including a deep-skirt block that retains the crankshaft with six main bearing cap bolts; tall-port, 15-degree valve-angle cylinder heads; a longer stroke and smaller bore; a redesigned oiling system; a composite (that is, plastic) intake manifold; and a distributorless ignition system.

One Size Doesn't Fit All

Every situation is a little different, whether because of your own abilities, your finances, or, of course, any limitations imposed by the class in which you choose to compete (or, laws, if you're sticking to trolling the streets looking for fun).

You'll have to take all that into account when planning what to modify or replace in your engine, and that's what the remainder of this book is designed to help you do: to understand your engine and your specific application well enough to handpick only what you need to do to attain the goals you set.

If there's only one nugget of advice that you take away from this chapter, let it be that you should always err on the side of caution. In other words, don't overdo things. Build your engine modestly, there will always be time for more radical steps later, if you need them.

CHAPTER TWO

SHORT-BLOCK ASSEMBLY

The short-block assembly consists of the engine block (sometimes referred to as a cylinder case) and the parts that fill it, except for the camshaft. In other words, you're dealing with the pistons, piston rings, connecting rods, crankshaft, bearings, and other miscellaneous components.

With the exception of a few changes here and there—different displacements, crankshaft journal sizes, rear main seal designs, and a few others—all tuned port injection (L98/LB2 and LB9) engines have a short-block assembly that is virtually unchanged from the very first Chevrolet small-block that was unveiled in 1955. In fact, the entire engine is quite similar, save for the induction system. That's good news, because it's part of what helps keep these engines affordable, since so many parts are interchangeable.

The LT1/LT4/L99 engine family—the so-called Gen II small-block—is much the same as the first-generation small-block, yet different enough that many parts will not interchange. For instance, the Gen-II crankshaft is almost entirely different up front, because of the Opti-Spark ignition system and different hub and harmonic damper system used. Likewise, the engine block is quite different, largely because of the Gen-II's reverse-flow cooling that requires large coolant passages in the front of the block. The front of the block is also machined quite differently, in order to mount the Opti-Spark ignition system and the reverse-flow coolant pump. Likewise, the lifter valley is machined differently from a Gen-I block. Since the Gen-II doesn't use a traditional distributor, a stub shaft is inserted to drive the oil pump, and a mounting boss is machined in the lifter valley to secure the shaft in place.

The short-block assembly makes up the bulk of your engine. In the foreground is a short-block assembly, while the engine in the rear is a "long-block" assembly because it has cylinder heads bolted to it.

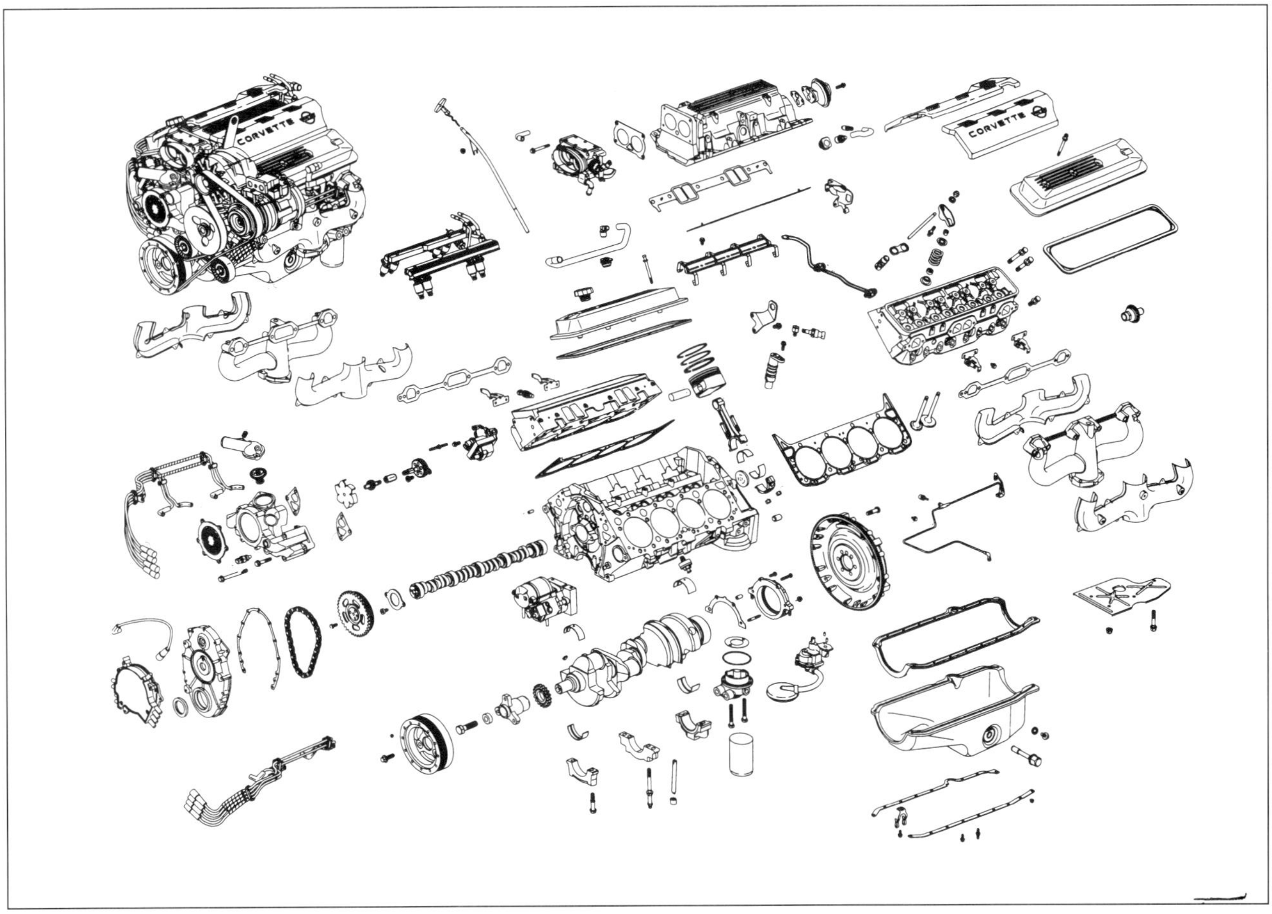

This exploded diagram shows the various parts that make up a Gen-II Chevrolet LT1 engine. Most of the components are parts of the short-block assembly. *Chevrolet*

The Gen-III small-block—the LS1—is a completely different animal from either of the previous generations of small-blocks. In fact, about the only things they share are a couple of dimensions, such as that tried-and-true 4.40-inch bore-spacing. Because the LS1 is so different, and relatively new (it debuted in the 1997 Corvette, in January 1997), we're not going to spend much time on the LS1 in this book. While it's true the principles of modifying one aren't profoundly different from those applied to a Gen-I or Gen-II small-block, the processes and parts used are. So, the subject of LS1 modifications is best left to a later date when more parts are available and more is known about the engine's capabilities and limitations.

The Gen-I/II engines, however, have been battle-tested on the streets, drag strips, race courses and, of course, the engine dynos. Much is known about just what these engines will handle, what they won't, and what you have to do to get the most out of them. In fact, part of what we know about these engines is that—depending on how you plan to use your TPI Gen-I/II engine—you may not even have to worry about modifying its short-block assembly. But since we've devoted an entire chapter to the subject, and since you can see lots more words on this page, you've probably already figured out there are plenty of things that can—and in many cases *should*—be done to improve your engine's short-block assembly.

According to Myron Cottrell of Cottrell Racing Engines and TPI Specialties, the factory TPI, LT1, and LT4 short-block assemblies will typically withstand up to 400 to 450 horsepower, because the original parts are purposely overengineered. GM does this for reliability and warranty reasons; building an engine with beefier-than-necessary parts ensures that few will fail under normal use. Consequently, the parts can tolerate moderate power increases with relative ease, especially on the street, where peak power is generated only occasionally, and almost never for more than a split second or two.

As proof, Cottrell has, himself, built countless 400-plus-horsepower TPI engines with stock short-block assemblies. Rather than spending time and money working on an already satisfactory short-block, he concentrates on the parts he bolts to the engine, carefully selecting and preparing them for maximum power gains.

It's especially fortunate that a stock bottom-end setup works just great for high-performance street applications, because few enthusiasts have either the equipment or the skills needed to work on the short-block assembly. Yet almost

The front of the engine block changed somewhat when the Gen-II small-blocks debuted in 1992. Of particular note is the water pump drive shaft core hole (top), the revised water pump passages, and the revised timing cover flange area. The center hole is where the cam goes, while the bottom hole is for the crankshaft.

anyone can bolt on different intakes or headers, or even swap cylinder heads, and the engine never has to leave the car's chassis (usually) for such jobs.

That said, though, there are indeed times when the stock short-block assembly just can't cut the mustard. For those times, there are plenty of improvements that can be made. Following are discussions of each of the major parts of the short-block assembly, and some of the various ways in which you can modify them, plus what you can expect to gain from them.

Engine Block
Selection

The engine block is the foundation of your engine, which means it's critical that it be sound and sturdy. Fortunately, most production blocks are more than suitable for high-performance use. With careful preparation and a few well-chosen improvements, they can survive even in the most demanding racing situations.

One of the biggest selling points for many enthusiasts is whether the engine block is a two-bolt or four-bolt block, referring to the number of fasteners that retain each of main bearing caps 2, 3, and 4. Four-bolt blocks are generally considered to be stronger than two-bolt blocks, but it has been shown time and again that two-bolt blocks are more than capable of retaining the crankshaft in engines that produce as much as 500 horsepower, especially if crank speeds rarely exceed 5,500 to 6,000 rpm for any length of time. The factory knew this and didn't waste time or money equipping production Camaro blocks with four-bolt mains until the introduction of the third-generation small-block, the LS1, in 1998, but then it had an excellent reason: The LS1 block is cast aluminum, which isn't as strong as cast iron, so engineers reinforced the block with not just four-main cap bolts but six! Four bolts run from the bottom of the caps up into the block, while two—one on each side—run through the sides of the block into the sides of the caps. Interestingly, the Corvette versions of Camaro engines—the Corvette's L98 (the equivalent of the Camaro B2L/L98 TPI 350) and LT1/LT4 engines all featured four-bolt main bearing caps. Apparently, GM figured Corvettes are likely to be driven harder than Camaros (though a case could definitely be made that in reality, the situation is just the opposite). So, if you insist on a factory four-bolt main block, hunt down a Corvette engine, many of which are easily found at large swap meets, like "Corvettes At Carlisle," or through Corvette publications, like *Corvette & Chevy Trader*. You'll also find that many Chevy and GMC truck engines are four-bolt main blocks.

But factory four-bolt mains aren't necessarily the hot setup. In fact, if you're building an engine for frequent competition use, you would actually be smarter to hunt for a two-bolt main engine block, because they can be modified to accept stronger four-bolt main caps than the factory's. Factory four-bolt blocks use four parallel bolts—each runs vertically

The center three main bearing caps in this block are retained by four bolts each. Lesser blocks use two bolts per cap for all the caps.

into the block. But with a two-bolt block, you can have it machined to accept four-bolt main bearing caps that feature splayed, or angled, outer bolts, which make the engine assembly much stronger. Note, however, that not all machine shops are capable of performing the needed processes to install splayed-bolt main caps. Check with the machine shops in your area if this process is appealing to you.

Another concern in selecting a block is its rear main bearing seal design. Small-block Chevy V-8s from 1986 and later use a one-piece seal, while earlier engines employ a two-piece seal. The primary advantage of the one-piece seal is its reduced likelihood for leaks and seepage, though with proper installation the two-piece seal can be quite reliable. While it wouldn't make sense to avoid a block simply because it has a two-piece seal, if you have a choice between otherwise identical one-piece or two-piece seal blocks, go with the one-piece seal, unless the engine is going to be used strictly for racing, in which case the availability of affordable aftermarket crankshafts far outweighs the inconvenience of a potential, minor oil leak.

Another difference between the two designs is that the one-piece seal cranks have a small-diameter flywheel bolt circle, compared to the two-piece seal crank. This can be important because flywheels for the larger bolt circle are marginally stronger and less likely to crack—or worse, to fail entirely—when forced to endure sustained high rpm.

All Chevrolet small-blocks made until 1987, including early TPI engines, used a two-piece rear main seal that was prone to leaks.

If you're building a first-generation small-block (one that has a distributor sticking through the rear of the intake), you will also need to concern yourself with the type of camshaft the block was designed for. Most early small-blocks used flat-tappet camshafts, but when Chevrolet began incorporating roller camshafts in the mid-1980s, the lifter valley machining was changed to accommodate the new roller lifters and the sheet metal "spider" that retains them. Older blocks can use roller cams with aftermarket roller lifters; however, factory roller cam blocks can use only roller cams.

Compare the two-piece seal design to this one-piece seal block. An aluminum seal housing is fastened to the rear of the block, filling the machined void along the rear edge of the block.

Also, as long as you're preparing a first-generation engine, you have the option of using aftermarket blocks designed for racing applications, including GM's own line of Bowtie blocks. These blocks permit larger than normal bore diameters and stroke lengths, plus can feature extra-tall deck surfaces, all of which allow for building mega-inch small-blocks. Additionally, Bowtie blocks are made of superior metal alloys (some are lightweight aluminum too), have additional material in critical areas such as main bearing webs, pan rails, and the lifter valley, and typically feature splayed-bolt (outer bolts only) main caps made of super-strong steel rather than cast or nodular iron. But all this added fortification comes with a hefty price tag, which usually precludes them from being used in anything except race machines where their superior strength may mean the difference between finishing a race first, or not finishing it at all.

Finally, you need to choose between a new block, or a seasoned

This is a close-up shot of the steel four-bolt main bearing caps with traditional straight outer bolts.

one—one that's already got several thousand miles on it. Believe it or not, this may actually be the most important decision to make, though for most people the smartest choice is the only realistic choice: a used block. That may sound contrary to what you may have expected, but used blocks have the advantage of having gone through the heating and cooling cycles of normal use countless times, which settles the block—relieves stresses locked in during the casting process. Obviously, however, you don't want a block that's been overheated, which can warp the block permanently. Likewise, any block that has suffered any kind of catastrophic failure—a broken rod, camshaft, or crankshaft, a piston failure, or what have you—may have caused irreparable damage to the block, rendering it useless. So be careful before shelling out money for a used block of unknown or suspicious origins.

Whatever block design you wind up considering—even if it's your original engine block that you're planning to use—it's vital that you perform the customary inspections. As a minimum, you should have the block cleaned, then measure the cylinder bores for taper and roundness, as well as for the presence of any gouges or a ridge at the top of the cylinders. Deck surfaces should be checked for flatness, and main bearing bores should be checked for roundness, size, and alignment. If you're building an all-out race engine, then you should also take the time to verify tappet (that is, lifter) bore placement, any core-shift that might have occurred, and other dimensions. Then, assuming it passes those tests, Magnaflux inspect it for cracks.

What's a "Stroker" Motor?

A popular and worthwhile consideration for an engine buildup is building a "stroker" motor—one with a longer-than-stock crankshaft stroke. While the stock 4.00-inch bore and 3.48-inch stroke combine to make a 350 cubic inch V-8, a long-stroke (3.75-inch) 400 crankshaft with a 0.030-inch cylinder over-bore produces a 383-inch small-block (377 cubic inches with a 4.00-inch bore) that produces more horsepower and torque than a 350 with the same effort. Note that this combination does require specific pistons with a reduced pin-to-piston deck height, to compensate for the increased travel of the piston, otherwise the piston would pop up out of the bore and contact the head!

Preparation

Because dirt and grime or even a light surface coat of rust can hide a lot, before you do anything with your block, clean it—thoroughly. Get a set of engine cleaning brushes from one of the mail-order engine supply shops and run them through every bore, every hole, every nook and every cranny possible. The best method is to have the block "hot

These caps feature splayed outer bolts, which make the cap much stronger and much less likely to "walk" from side to side. Also note that they are much wider than the straight four-bolt caps.

Here's a close-up of the underside of the LS1 block. Note that four vertical bolts plus two horizontal ones that pass through the skirt areas secure each main bearing cap. Also note the holes in each main bearing web, which bleed off cylinder pressures.

tanked," which involves a machine shop dunking it in a vat of super-strength cleansers. But even that isn't 100-percent effective, so still take time to manually go over and into every square millimeter of the block yourself. Grease-cutting soap, like dish soap, and hot water makes a good postprofessional cleaning solution, then dry everything off with compressed air, which makes it easy to blow out all the holes and corners.

After the cleaning, lightly coat any machined surfaces (the bores, decks, and so on) with a light oil, like WD-40, and wipe up any excess with a lint-free rag. This will prevent rust from forming on those areas.

Now, even though part of your block selection process was to closely inspect the part for excessive or abnormal wear or any signs of damage, this is so critical that it bears doing again. Recheck the block for any irregularities. Any scratches or shiny spots in the cylinders or any thrust surfaces should raise the hairs on your neck, because they indicate something isn't as it should be. It doesn't necessarily mean the block is useless—the damage could have been caused by a defective part that you removed from the block and won't be using again—but it would be wise to determine why or how the scratch or rubbed spot developed, so that you can correct the problem while rebuilding the engine.

If for some reason you didn't have the block Magnaflux inspected for cracks, do it now. This process involves applying a special fluid to the block (or other part), magnetizing it, then looking at it under a special light. When magnetized, the metallic particles in the fluid are attracted to cracks in the block, and those cracks become easily visible.

Magnuflux inspection only points out surface cracks, which for most people is usually enough. But for professional-level motorsports competition, simply magnafluxing a part is akin to judging a book by its cover. Pros need to know that a block is solid throughout—that the pages aren't blank or torn in that pretty-looking book. To do that, they often subject the block to sonic testing, which determines whether there are any voids (hollow spaces formed by trapped air bubbles) in the casting. Sonic testing also reveals any areas where the metal is thinner than it should be, due to a core shift during manufacturing or a poorly done overbore job.

This is also the time to break out the micrometers, dial-bore gauges,

One of the first steps of most high-performance buildups is boring the block. This one is being done with a torque plate installed, to simulate stresses of an installed cylinder head. This ensures the bore is perfectly round when the head is installed.

inside mics, and a straight edge to verify that every dimension of the block is within specs. The cylinder bores must be perpendicular to the crankshaft centerline, and should be properly spaced 4.40 inches apart, center-to-center. Check the deck height and whether the deck surfaces are flat and square to the crank axis. Measure the crankshaft centerline to camshaft centerline distance, and check that the saddles for both the crank and cam are in proper alignment with each other and are of the proper diameter. You also should check to make sure the lifter bores are properly oriented in relation to their corresponding cam lobe positions.

Machining

For 99.9 percent of the people rebuilding an engine, the major machining—boring, honing, decking, and so on—will be turned over to a professional, if only for the lack of proper equipment and knowledge of the procedures. But just because the actual work is out of your hands doesn't mean you don't have any control over what is done and how. It pays to know what machining processes are important and why, because it will not only help you determine what work your block needs, but also where you will have it done and by whom. Choosing a good machine shop is of major importance, so take the time to find a good one. (Here's a quick hint: Ask around at shows and race tracks to see who people recommend. The good shops will get lots of recommendations, and while that alone doesn't guarantee a good job, it's a good place to start.)

With all that said, the following information discusses the processes that you can have performed on your cylinder block, and whether you really need to consider spending the extra money to have certain processes done.

Boring and Honing the Cylinder

This is one of the most common processes to have done to a cylinder block. Whether the cylinders are scratched, out-of-round, have a ridge at their tops, or you simply want to increase displacement, boring and honing cylinders are pretty standard machining processes. Boring involves a cutting tool inserted in a boring bar to remove a significant amount of cylinder wall material in a short amount of time; it's the coarse file that gets things close to the final size that you want. Honing, on the other hand, is a grinding process that uses comparatively smooth stones to remove smaller amounts of material, allowing a machinist to get a block to an

Honing follows boring and should leave a distinct cross-hatch pattern in the cylinder walls, as shown here.

accurate desired dimension. The stones are mounted in a boring machine, which is essentially like a drill press.

Though it's hardly new, the Sunnen CK-10 is still the preferred boring machine, because of its accuracy and the superior cross-hatch finish that it leaves a cylinder with. It is possible to hone cylinders by hand; however, this is far from accurate and doesn't result in a consistent bore finish. Specifically, cylinder bores may end up out-of-round (egg-shaped) or tapered (fatter at the top or bottom, than at the opposite end), which limits how well the rings can seal against the cylinder walls. To improve the resulting cylinder bore even of blocks machined by a Sunnen machine, use of a torque plate is strongly recommended. A torque plate is simply a steel plate with holes in it that correspond with the cylinder bores, and with cylinder head bolt holes. It is bolted to the block during any boring and honing work, to simulate having a cylinder head installed, because when the heads are bolted on they distort the block slightly. If the cylinders are bored without a torque plate, they may end up perfectly round and straight (no taper) without the heads bolted on; but when the heads are bolted on, the bores may twist or bend out of shape, again reducing the ring seal.

Milling the Block's Deck Surfaces (i.e., Decking the Block)

The block's deck surfaces are the machined faces to which the cylinder heads mate, and there are three reasons to mill them: One, they may not be flat; two, they may not be parallel to the crankshaft centerline; or three, because you want to raise compression (by reducing the deck height). This process requires a special milling machine and cannot be done by hand, though it is essentially like planing a piece of warped wood to make it flat. It should be noted that as you plane the deck surfaces (or mill the heads' corresponding mating surfaces) you end up lowering the cylinder head intake ports, which requires milling the bottom of the intake manifold to restore proper port alignment.

If you are going to have your cylinders bored, you will then need to use oversize pistons and rings, because the stock pistons will then be too small for the new, larger bores. Pistons are widely available in +.030-inch and +.060-inch sizes, and you should order your pistons first, then have your piston diameters measured and add the desired clearance to that figure to determine the desired bore size or sizes (it's not uncommon for pistons to have slight differences in size, which could then call for subtle differences in bore dimensions).

Before you order pistons, however, get your machinist's advice on how far out your block will need to be bored to clean-up any irregularities, like ridges, scratches, gouges or worn, shiny spots. Then order correspondingly oversized pistons.

Align-Boring Main Saddles

Occasionally you'll find a block in which the main bearing bores don't line up

While boring and honing are best left to a machinist, almost anyone can debur their block. The lifter valley oil return holes are good places to start, since they usually feature considerable "flash" left over from the casting process.

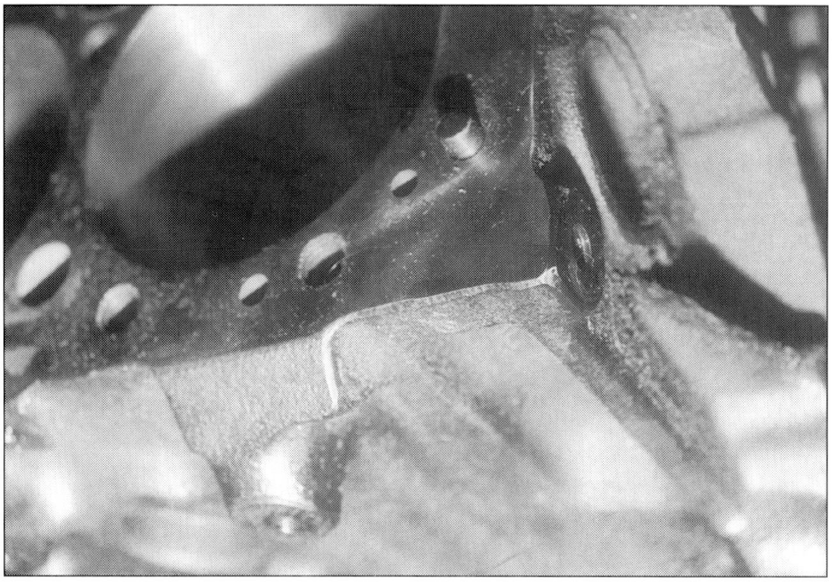

Another key part of deburring is relieving the block of any sharp edges that can serve as stress risers, which make cracking more likely. A quick pass with a grinder to bevel the edges is all it takes.

Another area that can be improved at home with a grinder is the oil passage entry in the rear main bearing cap. Note the rough as-cast finish and the hard transition into the oil gallery hole of this stock cap.

perfectly (at least, not within accepted tolerances). In those cases, the main bearing saddles are said to be in need of align-boring. In actuality, however, material shouldn't be removed from the block's main bearing saddles because doing so would physically move the crankshaft closer to the camshaft, and that would mean you wouldn't ever be able to get a timing chain set to fit properly (although special short chains are available, it's best to avoid the need for one if at all possible). Instead, the main bearing caps are removed, their mating surfaces are milled slightly, then they're reinstalled. Then the align-boring machine (again, this is not a "by-hand" process) removes material only from the caps, leaving the block's bearing saddles untouched, and thus the crank-to-cam distance unchanged. Align-boring is a required step if you ever replace a main bearing cap for some reason, such as when upgrading a block from two-bolt mains to splayed four-bolt mains via aftermarket main bearing caps.

Align-Boring Cam Bearing Saddles

Just as main bearing bores might not line up properly, so, too, can cam bearing bores be improperly located, either from incorrect machining or simply from settling of the block. In the case of align-boring cam bearings, machinists have to, again, take care not to reduce the cam-to-crank distance, but since the cam bearing bores are not two-piece assemblies like the mains, the cam bearing bores are drilled oversize, then special oversize bearings are installed.

Cam Bearing Installation

Cam bearings look like thin doughnuts that have to get pressed into their corresponding bores within the engine. The outside diameter of the bearing is slightly larger than the inside diameter of the bore, to provide an interference fit that will ensure the bearing stays in place. But that also makes them difficult to install. If you install the bearing slightly crooked, at the very least it will wear rapidly; at the very worst, you'll break the bearings trying to install them. This is a perfect job to leave to the professional machinists; if they break the bearings, they'll pay for the new ones. And if they install the bearings wrong, you'll at least have someone to point your finger at when the engine dies a premature death.

Tapping Bolt Holes

This is actually a controversial process. Many enthusiasts and machinists say you have to run a tap into every bolt hole in the block to ensure the threads are clean and burr-free for proper torque readings when installing bolts later on. But according to the folks at ARP—who make most of the high-performance fasteners used in every class of motorsports, including NASCAR Winston Cup stock car racing, NHRA Top Fuel drag racing, and SCCA Trans-Am competition—tapping a bolt hole removes material from the hole's threads, which means the bolt fits more loosely and the threads are weaker. Their recommendation then is to clean the holes as well as possible with engine-cleaning brushes, then tap only the holes that seem to need it and leave the others alone.

Clearancing (Pan Rails and So On)

If you happen to build a custom engine with a long-stroke crankshaft or aftermarket connecting rods, it may be necessary to provide added clearance for the crankshaft counterweights or the connecting rod bolts and caps. The most common areas to need added clearance are the block surfaces to which the oil pan mates (i.e., the oil pan rails). This is a job you can do yourself with either an air-powered or electrically powered die grinder and an assortment of grinding stones. The goal is to provide enough clearance that things won't ever come in contact with each other, without removing so much material that you weaken things or drill through to an internal oil or coolant passage. While we're on the subject of clearancing, we'll just mention that those oversize counterweights or connecting rods may also come in contact with the camshaft, too, which could require the use of a special small base circle cam, to provide adequate clearance up higher in the block. We'll get into this more in chapter 5.

Deburring

Deburring a block (or any part, for that matter) is simply the process of removing any sharp edges. Those sharp edges were either created when the block was being cast and excess iron oozed from the casting molds creating casting *flash*, or from the machining phases or preparation. Flash and sharp edges don't generally create many problems for regular passenger car engines, because they're so understressed most of the time. But sharp edges are actually *stress risers*—areas at which forces converge—and often result in the formation of cracks. Taking the time to gently round any sharp edges and grind off any excess casting flash (there's usually lots of flash in the lifter valley oil drain holes) with a die grinder can help prevent the block from cracking under high-performance use. It also has the added advantage of making the block easier to work on, since you'll be less likely to cut your fingers.

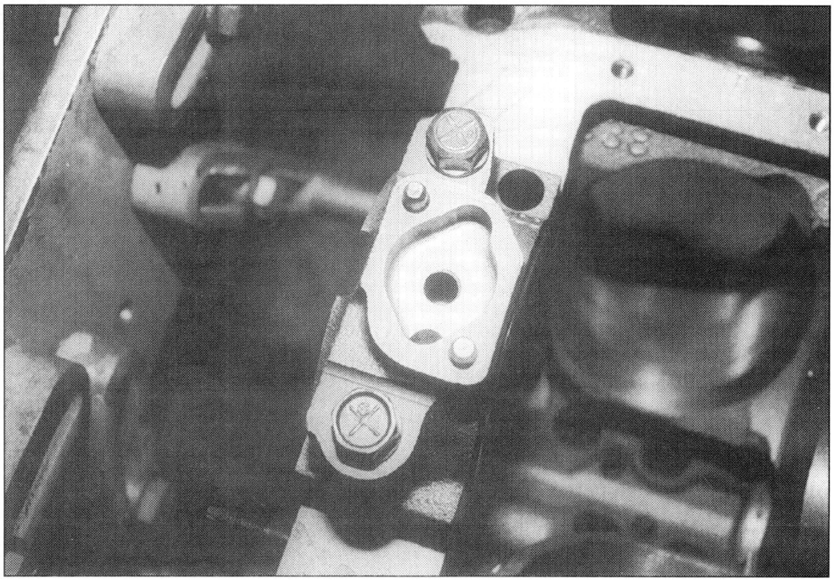

By comparison, note the smooth finish and contours of this CNC-machined cap. Yours needn't be this clean-looking, but smooth and with a gentle radius flowing into the passage will greatly aid oil flow.

Chamfering Bolt Holes

This is actually part of the deburring process, because it removes sharp edges, but it bears special mention because if deck surfaces have been milled it is especially important to ensure that the tops of the bolt holes in the deck surfaces are countersunk to prevent stress risers in this extremely critical area. A cone-shaped grinding stone is generally sufficient for this work, and don't get overzealous in grinding: A slight chamfer is all that's needed.

Cleaning (Again)

Once all the machine work is done, you need to clean off all the metal shavings. This is best done with very hot water, a good grease-cutting soap (like dish soap), and cleaning brushes. Again, dry, compressed air is ideal for drying the block and its many holes and crevices, and a light coat of WD-40 or similar lubricant on the machined surfaces will prevent rust.

Painting the Inside Surfaces

The interior surfaces of the engine that aren't machined (such as the walls and floor of the lifter valley) can benefit from being painted (such as with Rustoleum), which will speed the process of oil draining back to the pan, compared with the same oil having to ooze along the surfaces of the rough cast iron. Obviously, you would want to do this prior to oiling the block after cleaning and drying, to ensure the paint adheres to the iron well.

Converting Two-Bolt Mains to Four

This process can't be done by too many small-town shops because of the accuracy required, but it's worth mentioning. Blocks with main bearing caps that are retained by only two bolts can have their caps replaced with aftermarket four-bolt main bearing caps, which usually feature *splayed*, or angled, outer bolts for added strength. Factory four-bolt blocks can't be converted because their main bearing webs have already been drilled for four bolts. Heavy-duty Bowtie blocks do come from GM with splayed outer bolts. Because aftermarket main caps are used for the conversion, the block will also need to be align-bored to ensure that the crankshaft will spin freely in its bearings.

O-Ringing the Deck Surfaces

This is a process you're very likely not to need, because it really isn't necessary in anything but ultimate-performance situations, like NHRA Top Fuel drag-racing

Once the block is prepared, your attention should focus on the crankshaft. According to Herb Couzins of Callaway Cars, a crankshaft is one place where it really pays to get a high-quality piece. *Vibratech*

Like the block, the crankshaft's sharp edges should be beveled to reduce the likelihood of cracking.

motors. The process involves cutting square grooves in the block deck surfaces around each of the cylinder bores. Corresponding, perfectly aligned grooves *may* also be made in the mating surfaces of the cylinder heads, and when the engine is installed an O-ring is sandwiched into the grooves, virtually ensuring that cylinder pressure will never be lost due to a blown head gasket.

The Crankshaft

The crankshaft harnesses the forces pushing the pistons down in the bores and converts it to rotary motion that can be easily transmitted to the rear axle assembly and put to work turning the tires. In short, the crankshaft is the most essential component of a reciprocating engine assembly—it's what keeps things reciprocating.

Given the high loads placed on the crankshaft throws, it's essential that the crankshaft be in excellent condition; the slightest defect can lead to disastrous results—an expensive engine in pieces. But this doesn't mean you always need the most exotic crankshaft, you just need one that's well-matched to your driving conditions and is properly prepared.

Different Types

Factory production crankshafts have served countless drivers well for years and years, in both stock and modified engines, even under racing conditions. For the first-generation small-blocks, there are four crankshafts from which to choose: cast-iron or forged-steel units in either of the two rear main seal designs (one-piece or two-piece). Forged cranks are generally considered to be stronger, but they're also significantly more expensive to purchase, if you don't already have one. They can be identified by a wide (about 3/8-inch) parting line on the unmachined portions of the crank throws (cast-iron cranks have a narrow parting line, by comparison).

Cast-iron cranks, however, when properly machined and balanced, are perfectly suitable for high-performance use, as long as crank speeds are generally kept below 6,000 rpm.

The LT1 and LT4 Gen-II small-block V-8s all used the same crankshaft design—a forged-steel, one-piece unit machined for the LT1/LT4 Opti-Spark ignition system. Likewise, LS1 engines used the same crankshaft design, which was also a forged piece with the specific machining required by the LS1 engine design.

Aftermarket crankshafts are readily available in both cast iron and forged steel. The advantage of aftermarket cranks is the ability to get them in superior metals, with more precise machining, and in pretty much any stroke length you need.

Selection

The general requirements for crankshafts are that they, one, be crack-free, and two, are straight, meaning that the throws are properly indexed to each other and the crank isn't bent, which would prevent it from spinning freely in an engine. A competent machine shop can Magnuflux inspect the crank for surface cracks, and even sonic test it for internal problems if desired. A machine shop can also verify that the throws are properly indexed. As for straightness, a machinist would check it by laying the crank in a set off "vee" blocks and slowly spinning it with a dial indicator on each main journal; however, you can check straightness on your own using a slightly less accurate method that simply involves setting the crank in oiled bearing shells set in the block's main bearing saddles. Excessive movement at

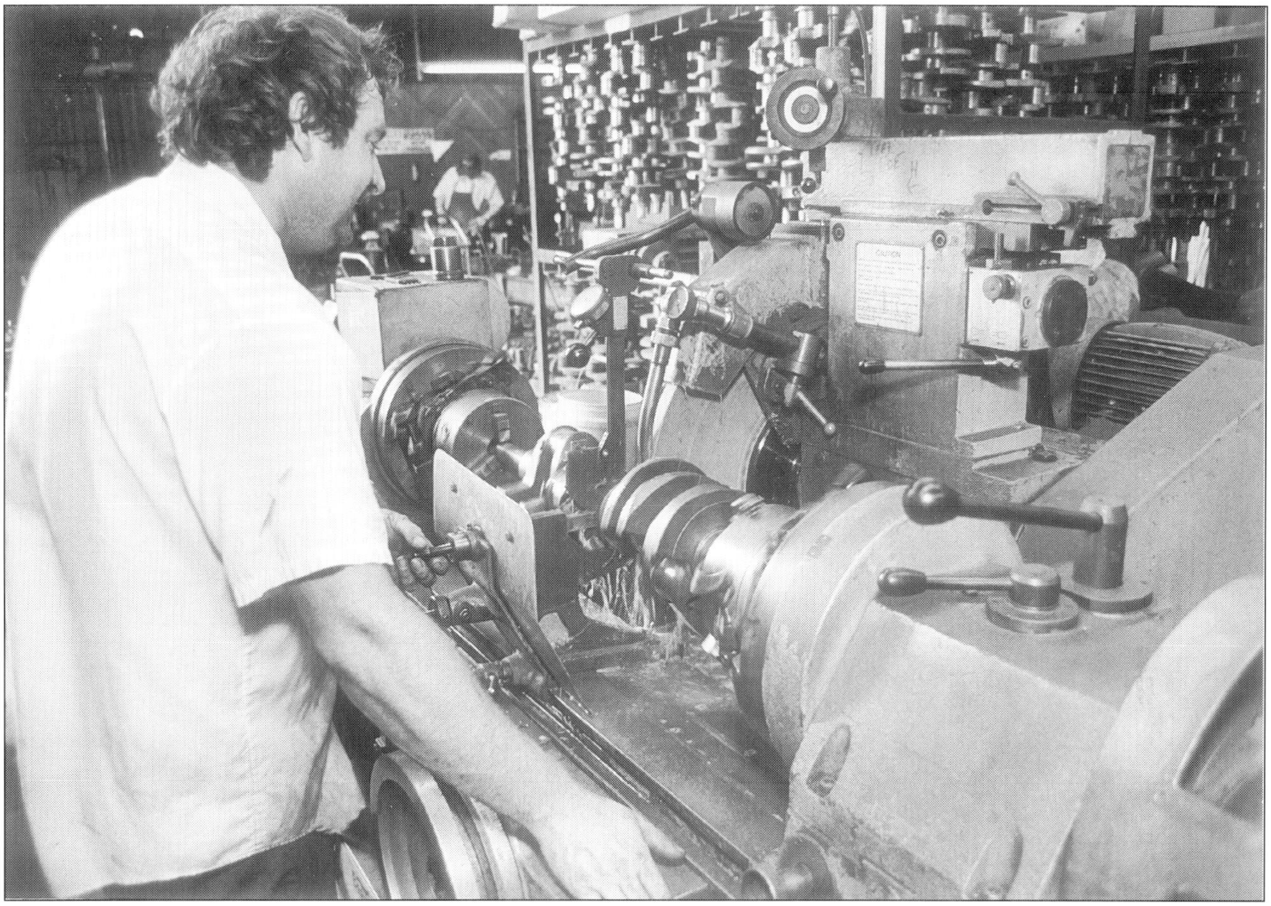

Preparation of a used crankshaft usually involves having the rod and main journals machined to a slightly smaller diameter to remove any imperfections.

one or more journals typically indicates the crank is bent (or at least that a journal is out-of-round).

Beyond those basic requirements, you might desire a crankshaft that is of a particular material—cast iron or forged steel, for example—of a particular stroke length, perhaps with a particular snout diameter (machined to accept either a small- or a big-block harmonic damper) or even that it feature lightweight hollow rod journals for quicker acceleration capabilities.

Naturally, whatever crank you are looking for must also fit the block you're planning to use, in terms of the rear main seal design: either two-piece or one-piece.

If you've got the choice between using a forged crank and a cast one, the forged would be the preferred option. If you lack the funds for a forged crank, however, don't feel like you're settling for inferior equipment; it's highly unlikely you'll ever suffer a crankshaft-related engine failure (other parts are far more likely to fail first).

Compare the first-generation small-block crank on the previous page with this Gen-III LS1 crank. The large notched discs are parts of the crank position sensing system.

Here's a close-up of the LS1 crank. Note the much larger journal diameters. The hole in the rod journal is only there for balance purposes and does not pass all the way through the rod journal; it does not dramatically lighten the crank's weight.

Machining and Preparation

Crankshafts are fairly simple—their only bolt holes are in the ends, up front to retain the accessory drive belt pulley, and in the back to retain the flywheel or flexplate—so preparation after proper measurements have been taken is largely in the hands of a machinist. Your only duties are really to make sure it is lightly oiled to prevent rust, and that it doesn't get damaged, dropped, or knocked over before it is secured in the block.

Once your crank is in the hands of a competent machinist, there are a number of procedures that he (or she) can perform on it to prepare it for high-performance use. As with most machining procedures, some are more important than others, and your crank may or may not need some of the work done, depending on its condition. But it's still helpful to know your options, so the following paragraphs describe the various processes and what they do.

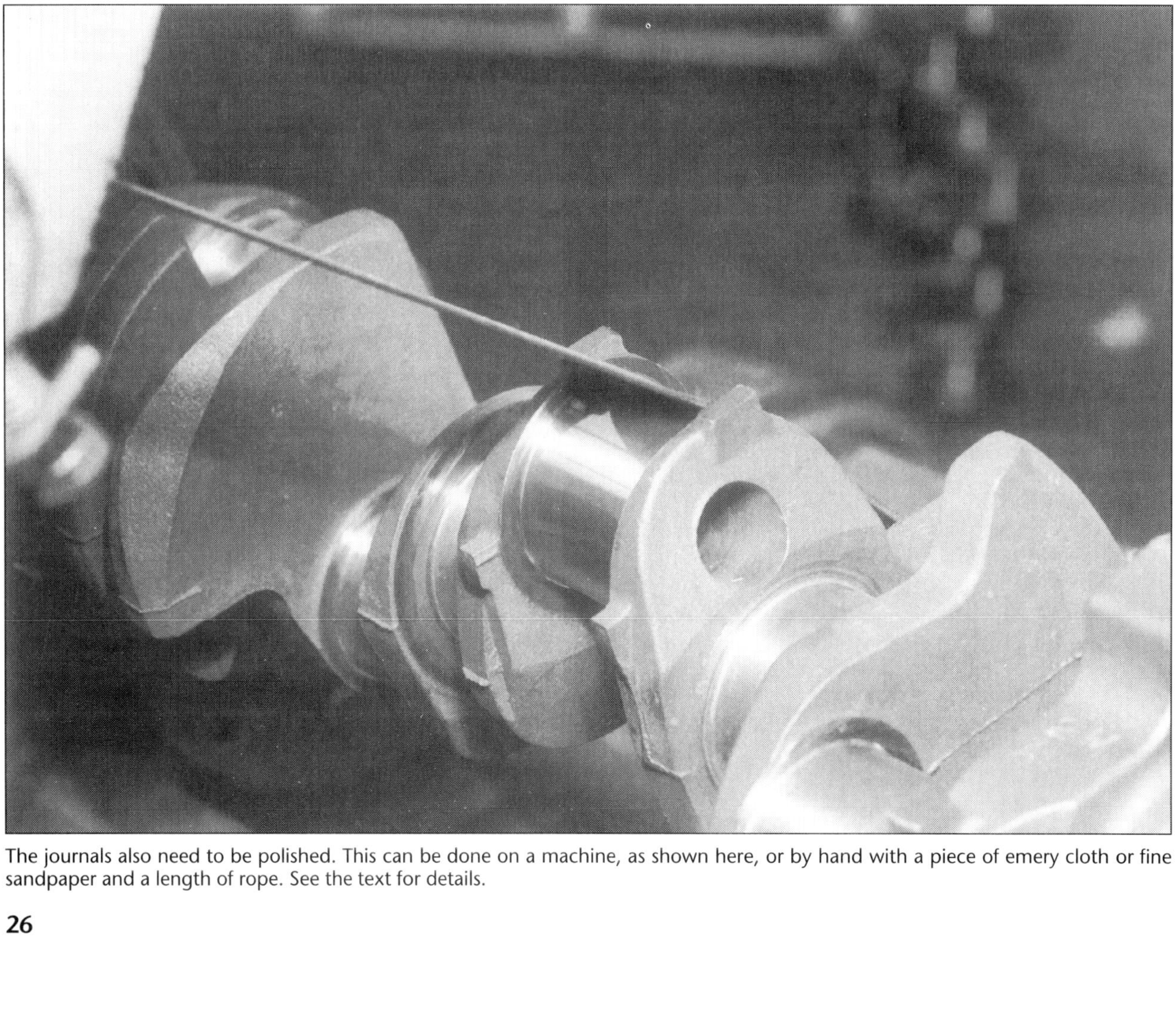

The journals also need to be polished. This can be done on a machine, as shown here, or by hand with a piece of emery cloth or fine sandpaper and a length of rope. See the text for details.

Magnaflux Inspection

Step one is to always confirm that the part is worth using, and that means looking for cracks and other defects or abnormalities. Magnaflux inspection is the usual route here, though racers can go to greater lengths (and expense) by sonic-testing the crank to reveal inner defects. Cracked cranks aren't worth much, unless, perhaps, you're handy with a welder and into industrial art.

Turning

If crankshaft rod or main journals are scratched, worn, or out-of-round, they can be restored by removing metal from them using a machining process referred to as turning the crank, until the journals are smooth and round again. Typically, cranks are turned in 0.010-inch increments, such as -0.010-inch, -0.020-inch, and so on. Oversize bearings must be used to compensate for the loss of crankshaft journal diameter, and you should realize that removing material from a crankshaft journal does weaken the journal, so you only want to have the minimal amount removed to clean up the journal. Because the surface material removed by turning is typically the hardened portion of the crank, the crank will typically need to be rehardened to restore strength.

Balancing

While balancing also involves the connecting rods, pistons, and rings—which are all statically balanced, using a scale—the crankshaft must be dynamically balanced while it's spinning. This is similar in concept to balancing a tire to have it spin properly and evenly. Not all machine shops can balance crankshafts, nor can all do it to the same degree of accuracy. For a high-performance engine, balancing should be considered an essential step, though, because the slightest imbalance can shake and pound an engine to death, flattening bearings (which increases clearances) and straining parts, possibly to the point of failure. If weight must be removed from a crank to balance it, request that the weight be removed by knife-edging the counterweights, rather than simply drilling them. (See knife-edging description, below, for more information.) Note that all small-block Chevy V-8s—with the exception of the early-1970s 400-cubic inch small-blocks, and the newer one-piece rear main seal engines—are internally balanced. There's also a difference between the 350 and 305 cranks, which makes the 350 crank far more preferable: The 305 crank isn't as heavy duty and often requires expensive heavy metal to be added to achieve a proper balance.

Note the large hole through the rod journal, which greatly reduces the weight of the journal, and aids in acceleration. This steel crank was manufactured by Callies.

This shot shows two important features: the rear journal design for a one-piece rear main seal and the holes drilled in the counterweight for balancing purposes.

Radiusing the Journals

Under the section on block machining, we discussed why it's important to debur the block: to get rid of sharp edges that can act as stress risers that can eventually lead to cracking. Crankshafts suffer the same sort of problem where the machined journal (rod or main) surface meets the throw surfaces, or "web," at a right angle. Typically, this junction is a tight inside corner. This area is referred to as a journal radius (even stock journals have a slight radius at these intersections, though they appear almost to be square corners). The larger the radius, the less likely a crack will form there. The only way to radius the journal edges, however, is to mill material away from the rest of the journal, leaving more at the edges, then carefully blending from the journal surface to the web. Interestingly, because of larger journal radii, a turned crank can actually be stronger than a nonturned crank, despite the removal of metal from the journals! This, by the way, is strictly a job for a qualified professional. Interestingly, the Gen-III LS1 crankshaft has virtually no radius, yet has super-strong journals thanks to a special surface-hardening process that compresses the journal in the normally radiused area, greatly increasing its strength.

Cross-Drilling

Another job for the professionals is cross-drilling the crankshaft journals to increase oil flow to the rod bearings. The term comes from the fact that the modified oil channels cross from one side of the journal to the other. Because of the accuracy required for this process, it's one that is best left to a qualified machinist who has the proper equipment for the job.

Chamfering Oil Holes

Another move that reduces the risk of cracking is to chamfer the edges of the oil holes in the journals. This can be done by the do-it-yourselfer, but be extremely careful not to damage the journal surface at all. When in doubt, leave this to the professionals. In addition to chamfering, according to General Motors' Motorsports Technology Group, oiling efficiency can be increased by grinding small (approximately 0.125-inch) grooves that begin about 1/2 inch before the oil hole (in the direction of crank rotation) and taper to a depth of approximately 0.125 inch at the hole.

Deburring

The oil holes aren't the only spots that need a little stress reduction. Removing casting flash and softening sharp edges on crank throws and counterweights will further reduce the risk of cracking, and it happens to be something you should be able to handle at home in an hour with a die grinder.

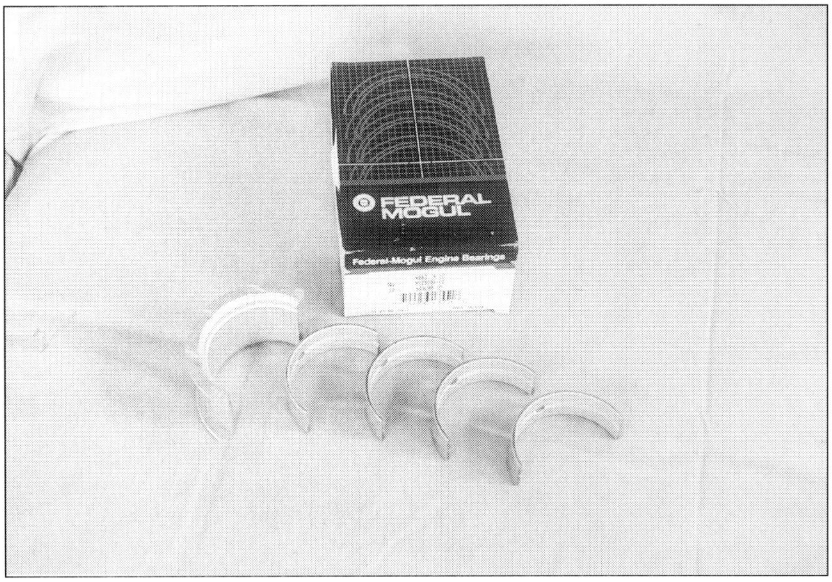

Good bearings are of critical importance to your engine. These halves are from Federal Mogul and feature a tri-metal composition that provides good durability and embedability to keep debris from damaging the crankshaft.

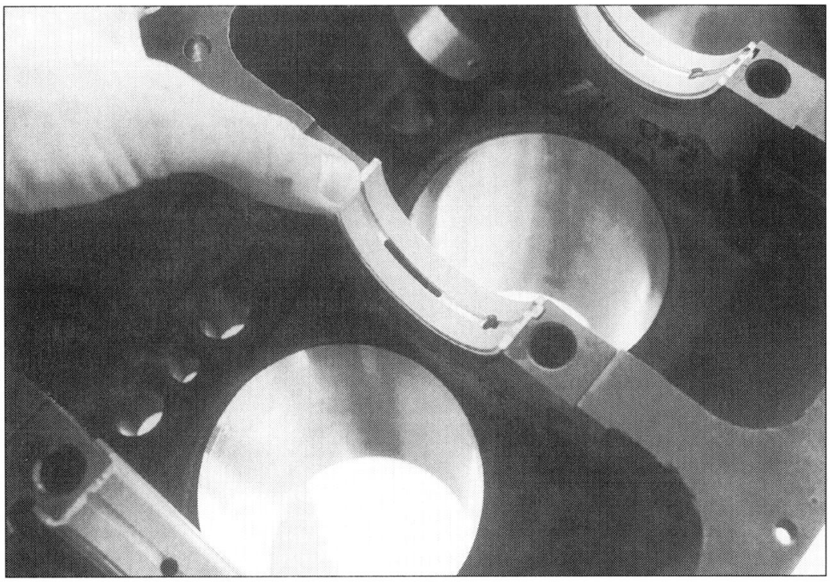

When installing the bearing shells in the block, make sure the oil holes line up with those in the block.

Beating the Bushings

For cars equipped with manual transmissions, you should always install a new pilot bushing in which the transmission's input shaft spins. An old pilot bushing may be slightly out-of-round and could induce chatter or excessive drivetrain noise, among other potential problems. GM offers a roller bearing–equipped pilot bearing (#14061685) that virtually eliminates drag, lasts longer, and is a direct replacement for a stock bronze bushing.

Knife-edging

If you've ever tried walking into a strong wind, you should be able to understand the benefits of knife-edging the crankshaft counterweights. Normally, counterweights are massive, slab-like hulks of iron (or steel) with sides that are nearly an inch across. Every time the crank rotates, the leading side of the counterweights smash through the air and oil in the crankcase and are slowed down, just like you are slowed by a strong wind. Furthermore, the oil—actually an oil mist in the air—presents an even greater resistance, much like you trying to run in water. But by radiusing the crank's counterweights so that they are rounded at the leading side, and reshaping their trailing edges to a near knife edge, the counterweight slices through the air with much less effort, so your engine doesn't waste power fighting itself.

Lightening

A bit more exotic than knife-edging is lightening a crankshaft, which is usually accomplished by drilling out the cores of the rod journals. The process is best left to a professional because of the machinery required to perform the process, and because it's extremely easy to destroy the crankshaft by exposing an oil passage. The primary advantage to lightening a crankshaft, of course, is that it requires less power to accelerate it, so more power is available to accelerate the vehicle. Note, however, that removing all this metal tends to weaken the crankshaft, which makes it inadvisable for street applications, where reliability should be more prized than ultimate performance.

Polishing

After a nodular iron crankshaft has been turned (material removed from its journals) the journal surfaces need to be polished while rotating the crank in a lathe in the direction it spins when the engine is operating (clockwise, as viewed from the front). This removes minuscule ridges that would otherwise shred the bearing's surfaces leading to premature failure.

Shot-Peening

Shot-peening involves blasting the surface of a part with tiny metal shot to pound down any slight imperfections.

In addition, as the surface is hammered by the shot, it strengthens it similar to how packing snow together makes a hard snowball. In the case of crankshafts, the parts you want peened are the journals' radii, so the majority of each journal should be masked. This is a job for the pros, not only because of the equipment required, but because it does take some skill to do it right.

Hardening

The goal of hardening crankshaft surfaces is to increase their strength so that they don't wear. This can be accomplished in many ways, including nitride heat treating, Tufftriding, or chroming. Tufftriding was a popular and effective method of surface hardening crankshafts during the muscle car era, and it involved giving the entire crankshaft a bath in a molten cyanide salt. The process was so effective that its 0.0001-inch penetration depth measurably increased surface hardness. Unfortunately, that cyanide salt bath isn't exactly friendly to the environment, so few companies offer the service these days.

Chroming involves protecting all but the crank's journals with wax (the journals are masked and the entire crank is dipped in wax, then the masks are removed and the journals cleaned), then the crank is dipped in an electroplating bath, after which the wax on the remainder of the crank is removed. The chrome is hard and smooth, but it can also flake off, which can cause tremendous damage to the engine.

The most common method of hardening a crank these days is to nitride heat treat it, which requires heating the crank to between roughly 900 and 1,000 degrees Fahrenheit (approximately 500 to 550 degrees Celsius) in a nitrogen-rich atmosphere, causing nitrogen to infuse the journal surface, hardening the surface metal to a depth of a few thousandths of an inch.

Note, however, that nitride heat-treating can only be performed on parts that have already been hardened by heat treatment, or the part may warp due to the heat. In any case, a crankshaft that has been nitride hardened should not be turned or machined, or the hardened surfaces will be removed, requiring it to be rehardened.

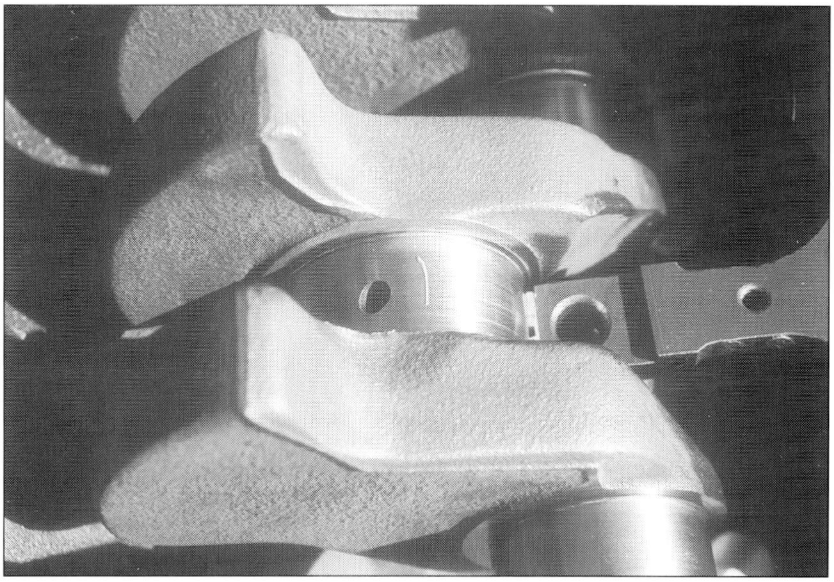

With the crank resting in the block, lay a strip of Plastigauge on a main journal, as shown, and be very careful not to turn the crankshaft at all . . .

. . . then torque all the main bearing caps in place.

Straightening

It's not entirely uncommon for a crankshaft's throws to be slightly out of phase with one another—perhaps 91 degrees or 89 degrees from other journals, rather than the ideal 90 degrees. This can happen during manufacture, machining, or even use. It is possible for some crankshafts to be straightened. Unless the crank is an especially rare piece, however, it is often best to simply locate and use a replacement crank.

Hand Polishing

This should be considered the final step of crankshaft machining, and it merely involves using a piece of 400-grit sandpaper and a piece of rope to polish the journals lightly. To do it, wrap a piece of sandpaper as wide as the journal completely around the journal with the abrasive side on the journal, and tape the loose end to the back of the sandpaper. Then wrap the piece of rope around the journal once, and alternate pulling each

A Few Words about Fasteners: Nuts, Bolts & Studs

Many expensive engines have been lost to the failure of a single bolt, nut, or other fastener. There are a number of reasons a fastener may fail, including its age, quality, or improper installation, to name a few. To minimize the chance of sustaining a fastener failure, you first should understand the forces that fasteners experience.

Basically, there are three forces bolts encounter: tension, shear, and double-shear. Tension forces are experienced when something tries to stretch the bolt along its long axis; nearly every bolt on an engine is subjected to tension forces, but main cap bolts are perfect examples: As the downward force exerted on the crankshaft causes it to try to separate the main bearing caps from the block, the bolts securing the cap to the block are stretched slightly.

Shear and double-shear forces load either one or both sides of a bolt, respectively, and are common on chassis components, like shock absorber mounting studs (shear) and control arm bolts (double-shear).

In addition to understanding the forces that act on bolts, it's important to understand how bolts and nuts work. Bolts (with or without nuts) are like rubber bands, as far as how they retain a part: Rubber bands hold two pieces together because the band is constantly trying to contract back to its natural state. That's precisely how bolts work: As they attempt to return to their at-rest condition, the tension between the bolt's threads and those of a nut or tapped hole create friction that prevents the bolt from loosening, and it holds the parts together.

When a bolt is overtightened, however, it may help to think of stretching a piece of taffy: Initially, the taffy resists stretching, but once it is pulled a short distance, pulling it farther becomes easier. Bolts react the same way; if tightened within their designed tolerance, the bolt will do its job well and will resist any further stretching. But if the bolt is overtightened, causing it to stretch beyond its designed limits, it will be forever weakened and on the verge of breaking.

Given this information, it should be clear why reusing fasteners can be so risky: They may have been stretched—either through improper installation or during use—beyond their useful range, and thus substantially weakened.

For this reason, you should never reuse connecting rod bolts. The forces they must endure are tremendous, and weakened bolts simply can't be expected to last.

Main bearing cap retaining bolts and cylinder head bolts, likewise, should be replaced with higher-strength fasteners. For mild performance street engines, new high-strength bolts are wise upgrades, but for hotter street buildups and all racing engines a better alternative is to use mounting studs instead of bolts.

Bolts Versus Studs

The difference between bolts and studs is in how each experiences loads, according to Robert Florine at ARP, which makes the fasteners used by virtually every top NASCAR, NHRA, and SCCA race team. While both bolts and studs are subjected to tension forces, bolts are also subjected to a twisting force, whereas studs are only subjected to the tension forces. Obviously, a bolt that is subjected to different forces is more stressed than a stud that is only loaded in one direction. In addition, studs can be secured to the block with Loctite fastener adhesive, which results in them becoming an integral part of the block, whereas bolts are not.

Proper Installation

ARP's Florine does not recommend tapping bolt holes because most bolt holes are already "sloppy." Removing more material by tapping the threads only makes the bolt (or stud) fit more loosely. Instead, Florine suggests a thorough cleaning with a small wire barrel-type engine-cleaning brush.

Most fasteners (bolts and studs) require some form of lubricant on their threads to provide accurate torque or stretch readings. New performance fasteners should include an installation instruction sheet that tells not only the lubricant to use, but also the recommended torque reading or the recommended bolt stretch.

In addition, some fasteners require special sealers or adhesives to ensure their successful use. Most cylinder head bolts protrude through the block's deck surface into the water jackets. To prevent seepage of coolant up the threads, the head bolts (or studs) need to be coated with sealant along the threads. As mentioned previously, if main bearing cap or cylinder head studs are being used, their threads should be treated with Loctite to keep them more securely fastened to the block.

All fasteners have a specified torque setting that prevents them from being overtightened. If a stretch dimension is specified, measure the stretch while installing the fasteners instead of the torque, because bolt stretch is a more accurate method of determining whether a fastener is properly installed.

end of the rope to rotate the sandpaper as quickly as you can, but only with light pressure.

Connecting Rods

Connecting rods link the pistons to the crankshaft. It sounds like a simple job, but it's actually incredibly demanding. Connecting rods must endure unimaginable forces that continually stretch and then

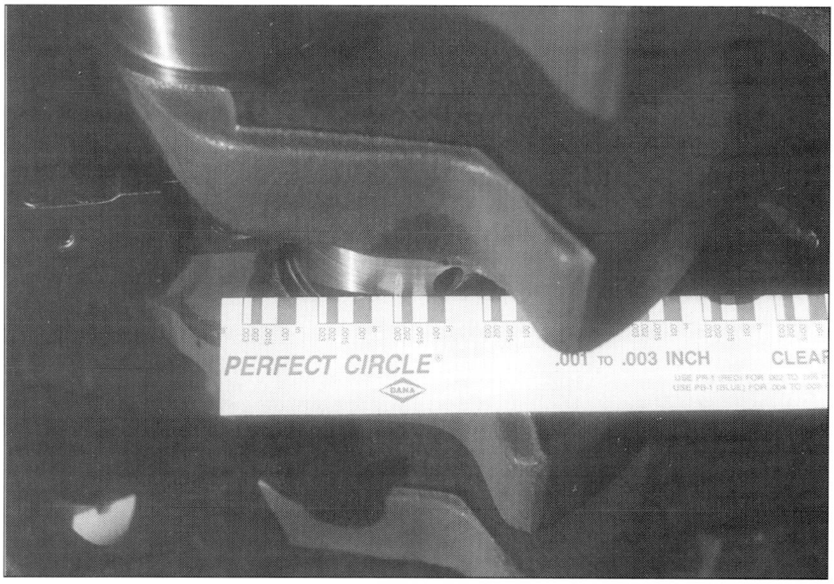

Remove the cap of the main bearing journal with the Plastigauge and compare the thickness of the squashed plastic to the measurements on the Plastigauge packaging to determine the oil clearance. Repeat these steps for each main bearing journal.

compress the rod during each revolution. As the rod pushes the piston up the bore, there are compression forces caused by the resistance of the piston and the speed with which the crank is thrusting the rod upward. As the piston reaches top dead center, the piston attempts to keep rising, but the rod must yank it back down as the crankshaft swings downward; this causes the rod to stretch. As the piston reaches bottom dead center, the opposite happens—the piston tries to keep dropping, but the crank starts pushing it back up, causing the rod to compress. Of course, the rod is also compressed by the combustion forces too. This process repeats thousands of times per minute and is especially punishing at high rpm, where the stretching and compressing forces are amplified.

Factory stock rods are excellent low-performance items and even work well in mild high-performance applications without much special attention. More extensive work can prepare stock rods for even severe service; stock rods have successfully been used in 9,000-rpm drag racing engines—and lived to tell about it. Of course, they did receive some special attention from the engine builder before doing so.

But stock rods aren't ideal performance pieces. Compared to most aftermarket rods, stock rods are heavy, which slows down engine acceleration, and therefore slows vehicle acceleration. Stock rods aren't made from the choicest of materials (1982–1992 rods are cast iron, while rods from 1993 and up are powdered metal, which offers similar strength to forged steel, but at a lower price). Forged-steel rods, such as the renowned Pink rods from the Z/28 302

After all the main journal oil clearances have been checked, you need to check crankshaft end play, or how much it can move front to rear. Pry the crankshaft forward with a large screwdriver then measure between the thrust surface and rear main bearing with a feeler gauge.

Connecting rods are the next area of concern. These are Competition Cams' Hi-Tech Billet rods, which offer superior strength and minimal preparation for high-performance and racing use. *Competition Cams*

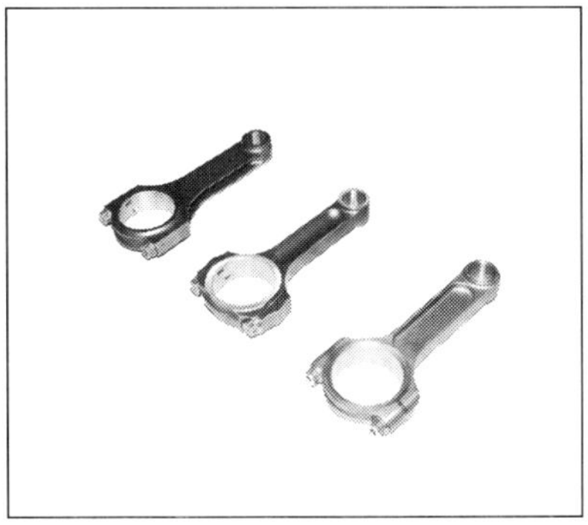

Connecting rods are available in many different shapes, sizes, and materials. These are three of literally dozens of possibilities.

and LT-1 350 are available through GM dealers' parts departments, and are good choices, but they're not inexpensive, nor are they lightweight. Nor are stock rods—whether of cast, powdered metal or forged construction—machined to precise tolerances; while the specified length for 305 and 350 V-8 rods is 5.700 inches, mass production tolerances allow for the length to be a few thousandths of an inch longer or shorter, so it's possible to have one rod that's 5.702-inches and another that's 5.698-inches in the same engine. The difference may not sound like much, but it can significantly affect engine operation, because the cylinder with the shorter rod will have less compression than the other cylinders, while the long rod will push the piston further into the chamber, raising the compression over that of the other cylinders. Furthermore, the cylinder with the highest compression will be more prone to detonate, requiring more conservative ignition timing, which will reduce power. In addition, different compression in each cylinder can cause an imbalance that can shorten engine life.

Aftermarket rods, on the other hand, have the luxury of being produced in far smaller numbers for enthusiasts who know their value, and are willing to pay for it. Typically, even inexpensive aftermarket connecting rods will feature specific design improvements over stock rods, such as larger, stronger rod bolts or even cap-screw-style bolts that allow manufacturers to greatly improve the strength of the big end of the rod. Additionally, they can be made of superior metal alloys, and can be treated to machining processes that are either too costly or time consuming for mass-produced rods.

Aftermarket rods can usually be specified in different lengths, which can greatly affect engine performance by altering rod angularity and piston acceleration. Longer rods tend to help an engine develop more power because the rod spends less time trying to push sideways and thus more time pushing the piston up or pulling it down as the crank continues its rotation. Plus, as Thomas Weiss at Callaway Cars pointed out, longer rods allow pistons to dwell at TDC longer, which keeps the air/fuel mixture compressed longer, and thus aids combustion because the mixture burns more completely. Shorter rods, however, tend to allow for quicker engine acceleration, but give up some power.

Not all connecting rods are made of iron (or steel). Aluminum rods are available for small-block Chevys (Gen-I and Gen-II engines, anyway), which can greatly reduce reciprocating mass, allowing for more rapid engine acceleration. Aluminum rods aren't suitable for use in street engines, however, because of the significant expansion rate of the metal and the rate at which it fatigues. Professional drag racers love aluminum connecting rods, and aluminum connecting rod manufacturers love professional drag racers, because the racers replace their rods after a very limited number of runs down the strip.

Selection

If you've got a 1993 or newer LT1 or a 1996 LT4, the powdered metal rods are likely to be just fine for high-performance street applications, when treated to the proper machining and preparation steps. If you haven't already got powdered metal rods, or you're planning to build some serious horsepower and torque with your engine, then you would

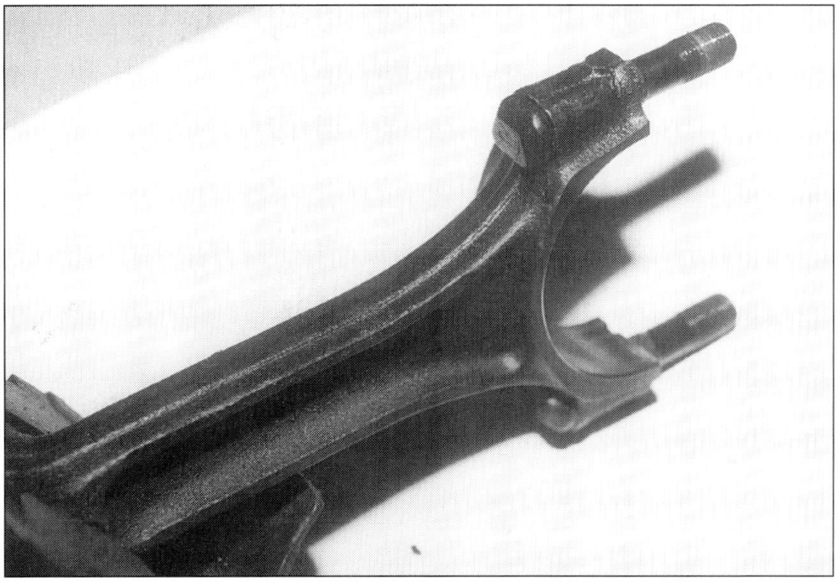

Stock rods, though hardly exciting or ideal, can serve you well, even in mild performance small-blocks, if prepared properly. One of the most worthwhile modifications is using a die grinder to remove the large parting line on the side of the beam, shown here. Also thoroughly debur the rods to lessen the chance of cracks forming.

be wise to consider investing in a set of forged-steel aftermarket rods. Stock cast-iron rods will work just fine for moderate high-performance engines, but the money you'll invest in them to prepare them will be nearly equal that of purchasing a set of ready-to-use aftermarket forged pieces. (The same could also be said of reusing powdered metal rods.)

For the record, it's worth noting that buying connecting rods at a swap meet is a risky proposition. There are a number of defects that are undetectable to the naked eye that could mean a particular rod or set of rods is unusable, meaning that you would have wasted money. Among other things, the rods could be bent or twisted, stretched, have ovoid-shaped journals (at either end), or, of course, they could be cracked. At the very least, some portion of the rods' reasonable life expectancy will be used up. Even if the set you get isn't damaged beyond repair, you'll usually have to perform the customary rebuilding processes, and that means, in the end, it probably would have been smarter and cheaper to simply order a new set of aftermarket rods that don't require any preparation.

Whatever rods you do wish to work with, inspect them carefully for weak areas, since not all rods are machined identically, and some are definitely superior to others. For instance, pay particular attention to the sides of the rod at the small end; it's fairly common for the small end bore to be machined off-center, resulting in one side being thinner than the other. That thin side may be too thin to be relied upon. Likewise, the same can be true of the big end, or if the rods have been rebuilt once or twice before they may have relatively little material left in the rod cap, thus compromising strength.

You also need to be leery of rods that have anything but the faintest signs of rust on them. Rust attacks the metal, causing pits, and those pits weaken the rod. While sand-blasting the rods may remove the rust, it won't remove the pits, and it may actually make them worse.

Machining and Preparation

If you're rebuilding your short-block assembly, there are some processes to which you can subject your connecting rods that will strengthen them and improve their reliability.

Magnaflux Inspection

As with many other parts, the most basic step is to have the rods Magnafluxed for crack detection. Cracked rods make excellent paperweights. (If you get enough of them, they're even decent wind chimes.)

Resizing

To ensure that both the small and big ends of the rod have the intended dimensions, and that the bores are round, rods get resized. On the small end, that may entail reaming out the bore and installing a bushing that reduces the opening to the

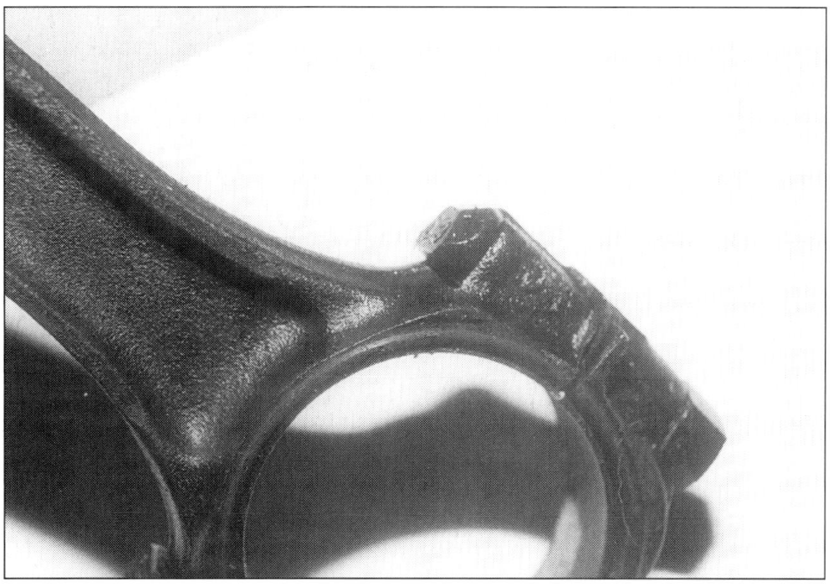

One area that requires particular attention on stock rods is the small shoulder against which the head of the rod bolts contacts. Remove the bolts, then bevel the shoulder to lessen the chance of a crack forming in this highly stressed—and thin!—area.

Engine Kits

One of the easiest, quickest, and most economical ways to rebuild your engine is to purchase an engine rebuild kit that includes the bulk of the necessary parts, from pistons, rings, and bearings to gaskets, core plugs, and even an oil pump. Some more specialized kits may include reconditioned or new connecting rods and even a crankshaft in your choice of materials, strokes, and journal sizes. At the high end of the engine kit market are the complete kits that include everything—even the cylinder block.

While an engine kit reduces your chance of forgetting something, they also tend to limit some of your options, and it's unlikely that all the parts will be precisely what you would have chosen if you were to select each individual component. Often, the parts in kits are not the ultimate high-performance piece, but rather a middle-of-the-road component that will be durable and provide good performance at a reasonable price.

You should also be wary of kits that consist of "no-name" parts. While it is possible that such parts are manufactured by name-brand companies like TRW, Sealed Power, or others for stores to "brand" as their own, it's also possible that the parts are poor-quality replacement parts manufactured by unknown companies. If the brands of the parts included in a kit aren't clearly stated, take the time to ask who manufactures them and always find out what the guarantee, warranty, and return or exchange policies are.

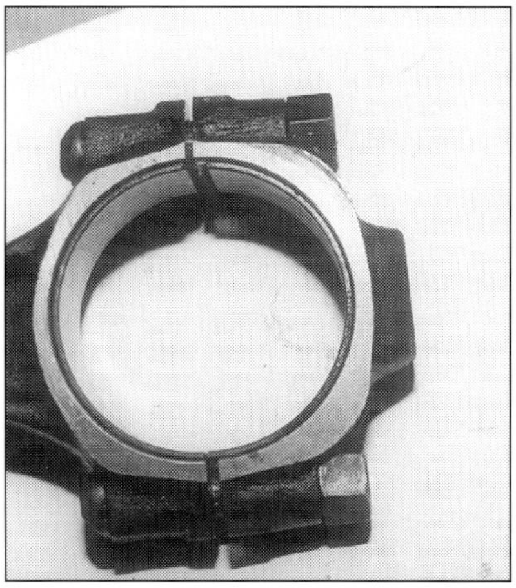

Stock rods typically have substantial "balance pads" at each end. These are where to remove material for balancing purposes; however, do not assume that you can simply grind away the pads entirely. Doing so may actually create balance problems.

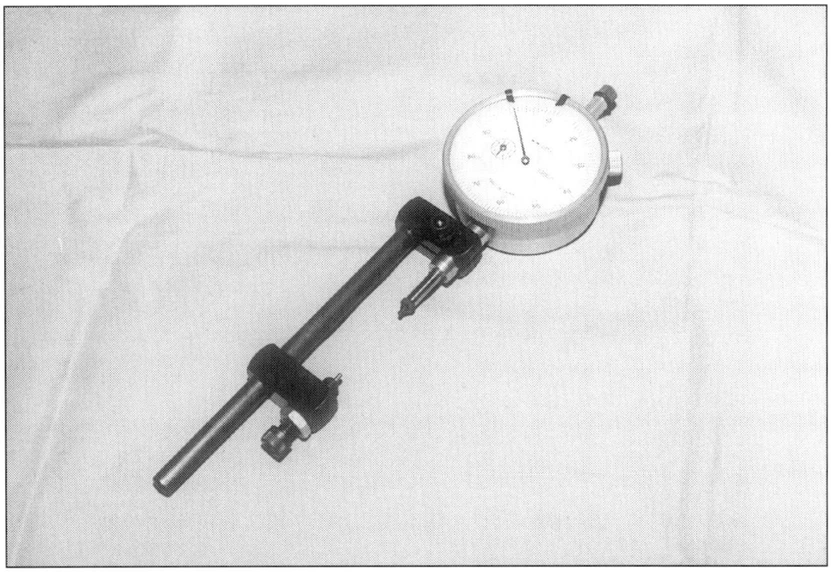

When installing rod bolts, it is preferable to measure their amount of stretch (as opposed to simply torquing them in place). This is a simple stretch gauge made just for that duty.

proper size, or better yet (for high-performance use, anyway) you could rework the rods so they have a floating pin design, which is actually the same process, but the bushing allows for a slight clearance. Meanwhile, the big end can have its cap milled slightly, then the journal bore re-bored to the proper size.

Deburring/Polishing

Next, the rods should be deburred and preferably polished to remove any stress risers that could lead to cracking, particularly along the sides of the rod beam (the long, skinny section between the small and big ends), and at the rod bolt interference notch, where the head of the rod bolt contacts the rod. It is especially important to radius the rod bolt head notch area, because the area is thinner than most performance engine builders would like it, but can't be enlarged; by radiusing the area and minimize the shoulder against which the side of the rod bolt's head contacts (to help prevent the bolt from spinning in the rod), it actually decreases the likelihood that the area will be overstressed. The interference fit of the rod bolt into the rod should be sufficient to prevent it from turning in the rod while the bolt's nut is torqued. As for the rest of the rod, when deburring and polishing, it is best to file along the rod beam from end to end, rather than from side to side, to minimize the chance that you'll actually create transverse scratches that could act as stress risers.

Shot-Peening

Shot-peening the rods' exterior surfaces provides the same benefit it does with the crank's journals: It hardens them and removes minor surface imperfections. Most machine shops consider this part of a standard reconditioning job, but ask yours just to make sure.

Rod Bolt Replacement

You should never reuse old rod bolts, especially after resizing the big end of a rod. Replacing the stock rod bolts with

Pistons are often tops on everyone's replacement list in a performance rebuild, but there are many different types available, and some preparation that they deserve. This is a Sealed Power .030-inch-over, four-valve relief, forged flat-top slug with its matching piston pin. The dot is to be placed toward the front of the motor to properly position the thrust side of the piston.

This is a Lunati forged aluminum flat-top piston with two valve reliefs. Pistons should be thoroughly deburred prior to use too.

Another option is a dished piston, like this cast unit.

high-performance rod bolts will greatly improve the strength of the rod assembly. Traditionally, bolts with knurled shanks were the best choice for rebuilds, but Automotive Racing Products' (ARPs) wave-lock bolt design, which debuted in the early-1990s, has gradually proved itself superior and is now the top pick of professional race engine builders in NASCAR, NHRA, SCCA, and other forms of motorsports. Instead of knurled shanks, the ARP Wave-Loc bolts have a series of rounded humps that grip the inside of the rod without damaging it as knurled bolts do.

It's worth noting, here, that because of the way rod bolts are loaded and the forces they're subjected to, it is extremely critical that they be tightened properly. The thought of torquing them in place naturally comes to mind, and that's a better-than-nothing solution. But the best method, according to professional engine builders, is to measure the stretch of each rod bolt while tightening it. The reason for this is that fasteners have a certain degree of elasticity designed into them—that's what keeps them tight. If they aren't tightened enough, the cap will move—perhaps ever so slightly—which will eventually cause the bolt to fail. Conversely, if you stretch the bolt too far, just like an over-extended rubber band, the bolt may snap. In addition, bolts that have been overstretched (but didn't break) are usually weakened so significantly that they shouldn't be used. For a simple comparison, you might think of rubber balloons that you have to stretch first in order to be able to inflate them. Once they've been stretched, they become easier to stretch; rod bolts act the same way—except they often break after being overstretched.

Lightening and Balancing

Finally, the rods should be balanced, which simply requires finding the lightest rod in a set and grinding material from the heavier rods to make them equal the weight of the light rod, sort of. It's actually a bit more complex than that, because ideally the rods should be balanced end-for-end, meaning that every small end should weigh nearly the same, and every big end should weigh nearly the same. This can be a challenge, and grinding has to be done slowly, removing only a little metal at a time (and preferably repolishing after) to make sure you don't lighten any one rod too much, which would require you to regrind all the others to equal it. The only acceptable places to lighten rods are the machining pads on the big and little ends; that's what those pads are there for, and if you remove metal from anyplace else, you're likely to compromise rod strength.

Pistons and Wrist Pins

Pistons have the unfortunate duty of compressing the air/fuel mixture, then standing up to the intense heat and pressure that result from that mixture burning after ignition. The pistons may even have to endure the occasional detonation episode, which can be extremely damaging. And they have to do this thousands of times per minute! It's a cruel existence, to be sure.

Pistons come in three basic flavors: cast aluminum, hypereutectic (cast with

Domed pistons aren't as desirable because they inhibit travel of the flame front during the combustion process.

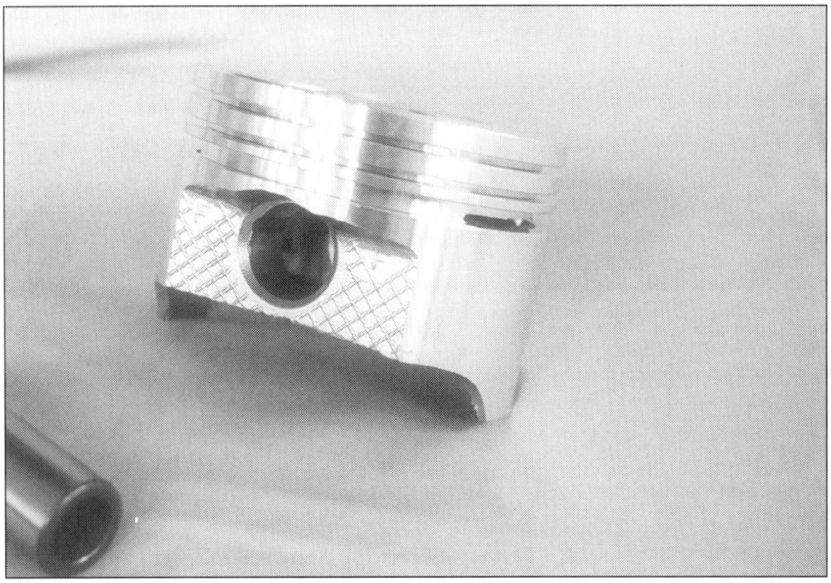

This side view of the Sealed Power piston clearly shows the three-ring grooves and the piston pin bore.

a high-silica content), and forged aluminum. Cast pistons are strictly low-performance items and aren't suited to high compression, nor will they withstand sustained high rpm (more than 5,500 rpm); but they are very stable, in terms of size, and they run very quietly. Most production engines used cast pistons because they were reliable and far less expensive than forged pistons.

Hypereutectic pistons are sort of a middle ground between cast and forged pistons. Made of a high-silica content aluminum alloy, they essentially combine the advantages of cast aluminum (stable size, reliability, cost-effectiveness) with increased strength, which had always been the big advantage of forged pistons. In recent years, General Motors has begun using hypereutectic pistons in a number of its production and over-the-counter engine assemblies, because they're essentially a best-of-both-worlds solution.

Forged aluminum pistons have long been the choice for high-performance and racing engines, mainly because of their strength, and they still hold a significant advantage in this regard over pistons made by other processes. But forged aluminum pistons are expensive, and, because they expand a great deal when subjected to heat, they must be installed with fat clearances that allow the piston to move around in the bore when cold, and thus are rather noisy (the noise typically fades as the pistons heat up) and can lead to accelerated wear. But forged pistons can handle higher temperatures, higher loads, and higher engine speeds, which keeps them the top choice for racing use.

The wrist pins (also called piston pins) connect pistons to the connecting rods, and raise another aspect of piston design worth thinking about: the pin type. Some pistons use pressed-in wrist pins and some use floating wrist pins. Pressed-in pins may be pressed into the piston but left floating in the rod's small end, or they may be floating in the piston but pressed into the rod's small end. Floating pins float in both the piston and rod and are typically used only in professional competition engines. For most applications, pins that press into the rod but float in the piston are the best choice. Pins that float in the piston and rod require that the piston pin bores be machined to receive snap-ring locks to keep the pin from sliding out of the piston and possibly contacting the cylinder wall.

Selection

In most cases, when rebuilding an engine you're probably going to use aftermarket pistons, simply because they're available in different compression ratios, different overbore sizes, and typically cost less than factory parts. Considering that you're even reading this book, chances are pretty good you're more interested in the strength of forged pistons than the value of cast or hypereutectic pistons. Whatever material you choose, you should pay particular attention to the weight of the pistons you're buying. Factory pistons are notoriously heavier than necessary to minimize the chance of failure, but the extra material also tends to slow down engine acceleration. Naturally, you'll also need to be concerned with compression ratio, which also requires taking into account the combustion chamber volume of your cylinder heads, the thickness of the head gaskets, the camshaft profile, the fuel you'll be running, and even the engine design itself. Gen-I Chevy small-blocks that are forced to run on pump gas (92-93 octane) won't react terribly well to compression ratios over 9.5:1. Beyond that the engine will detonate under all but the lightest loads, causing the knock sensor (if equipped) to retard timing so drastically that performance will be noticeably hampered. Engines without a knock sensor will simply detonate like crazy, potentially destroying the engine.

At least partially because of their LT1/LT4/L99 Gen-II reverse-flow cooling system, engines are more tolerant of higher compression, since the cylinder heads run cooler. Compression ratios of 10.5:1 are easily acceptable, and even 11.0:1 can be tolerated (though performance may vary depending on fuel quality and other factors).

The LS1 engine was still too new when this book was produced to have an assortment of aftermarket pistons readily available, but piston manufacturers are making sets, which to this point have largely only been used by the aftermarket tuners, such as Callaway Cars, Lingenfelter Performance Engineering, and Doug Rippie Motorsports.

A final word on piston design: When possible, it's best to use a flat-top or even dished piston, because pop-up piston domes can interfere with the travel of the flame front during the combustion process.

The job of selecting piston pins is often an easy one: Pistons typically include an appropriate pin. Of course, most piston manufacturers will happily sell you (or substitute) the specific pins you want. High-strength piston pins are available, as are lightweight piston pins. High-strength pins are a good idea, even if they weigh a few more grams than a standard pin. Lightweight pins, though, should only be considered for use in competition engines that will have them frequently replaced.

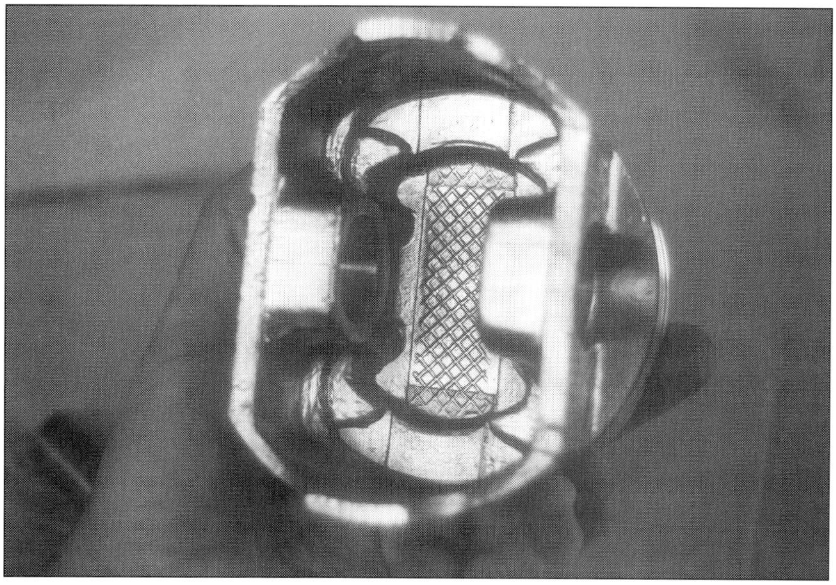

Inexpensive pistons often have considerable excess material underneath them, which causes them to be quite heavy, which inhibits acceleration.

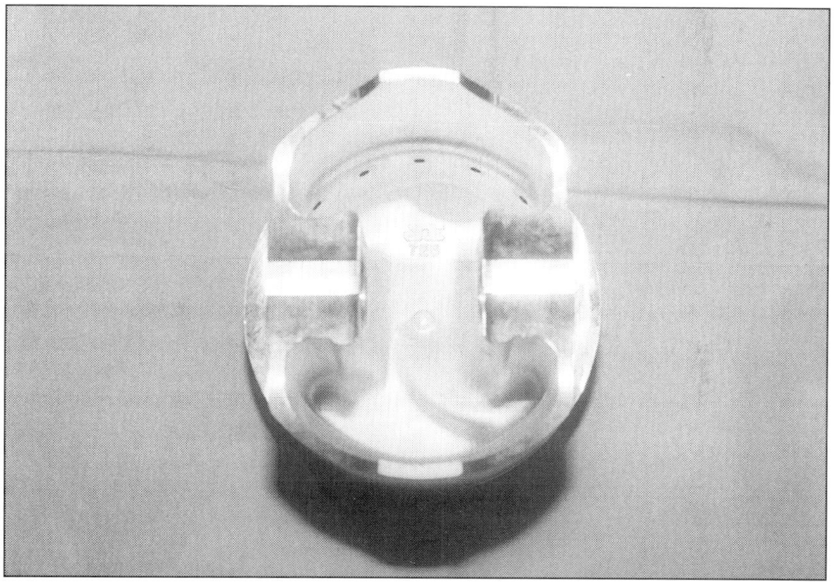

By comparison, the underneath of this JE Pistons piston is a work of art and features little excess material, which contributes greatly to its light weight.

Machining and Preparation

Just like any other engine component, pistons and pins need to be prepared properly if they're going to last for thousands of miles. Used pistons should be thoroughly cleaned, including the ring grooves, which will require a ring groove cleaner that scrapes carbon buildup and other deposits from the grooves.

All pistons should be checked for cracks by having them Magnafluxed, then they should be deburred to remove any potential hot spots. Finally, piston and pin assemblies should be balanced to match the heavier assemblies to the lightest assembly's weight. And if you're really into competition use, unless your pistons are lightweight units to begin with, you may want to consider shedding a few grams here or there in the name of quicker engine—and ultimately vehicle—acceleration. While it's best to consult with your piston manufacturer to determine what parts of the piston can be safely machined, in general you'll be

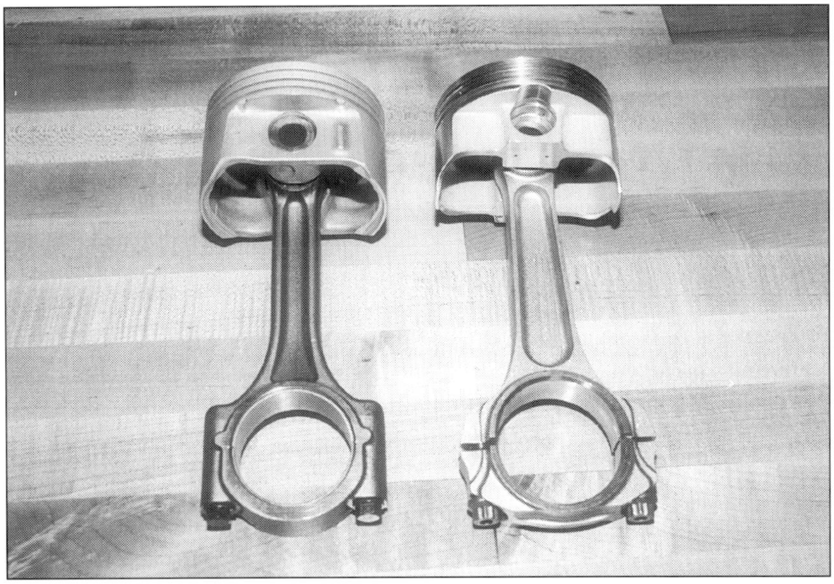

On the left is the stock LS1 piston and connecting rod assembly. Despite the beefy look of both the piston and rod, neither is extraordinarily strong. The piston and rod on the right are the Callaway Cars–designed replacements used in the SuperNatural LS1. The Callaway rod is longer and stronger, while the piston is lighter and stronger.

looking at grinding any casting flash on the underside of the piston, beneath the dome, as well as the piston pin bosses.

Piston Rings

Pistons don't completely fill the cylinder bore, because they need some clearance to allow for expansion and an oil film to lubricate the cylinder walls. Of course, if this clearance gap wasn't filled, the engine would hardly develop any compression at all, and therefore it would hardly make any power. Plus, it would burn oil and the oil would quickly become contaminated by gasoline.

Preventing all that is up to the piston rings. Piston rings are little more than seals that fit in grooves on the piston and help seal the piston to the cylinder walls. Most engines use three rings per piston. The top ring is the compression ring, and it is intended to help prevent the compressed air/fuel mixture from blowing out of the cylinders, plus they ensure that the rapid expansion of gases caused by the combustion process isn't lost in the same manner. The second rings help further limit any pressure losses. The lowest ring groove is occupied by the oil control ring package, which does as its name implies: controls oil, limiting how much is allowed above it.

Most piston rings don't make a perfect seal because of the ring end gap—the point where the two ring ends come together. Because the rings expand as they get hot, there is, by necessity, a slight gap that allows the rings to expand without breaking. Generally speaking, rings are sized to fit a particular bore diameter (such as 4.030 inches, for a 0.030-inch overbored 5.7-liter Chevy), and you'll need to verify that the ring gaps are within accepted tolerances prior to assembly (square the rings in their respective bores and measure the gap with a feeler gauge). If you desire a specific gap, you may need to order rings that are slightly oversize, to allow you to file one end of the ring to achieve that specific desired end gap.

Selection

Essentially, you need to be concerned with three things when selecting piston rings: the ring material, including any coatings that may be applied; the ring sizing and end gaps; and the oil control ring package.

In the old days, the top two rings were typically just plain iron, but performance top rings are now typically moly-faced or plasma-coated to improve their sealing and wear properties. Second rings still are typically plain iron, while the oil control package usually consists of steel upper and lower scraper rings, plus a steel expansion ring between the two.

As we covered in the previous section, piston rings require a gap to allow for expansion when the rings heat up; however, the larger the gap, the more compression you'll lose, so end gap is a critical setting. Certain piston ring

The piston on the left is Callaway's forged replacement used in its SuperNatural LS1. Because of a higher compression height and taller camshaft, valve reliefs are required. The piston on the right is the stock LS1 hypereutectic unit.

Pistons aren't any good without piston rings. Though they're referred to as a three-ring set, there are actually five pieces—the top (compression) and second rings, plus the oil control package of two scrapers and an expander.

designs actually provide a zero gap by having the ring ends overlap, sort of like tongue-and-groove boards, rather than butting against each other. Zero-gap rings are the ultimate in cylinder sealing, because they virtually eliminate blow-by, thereby maximizing power.

Oil control packages are available in a special low-tension design to reduce friction, but be wary of such a design. It can result in excessive oil consumption and unsightly puffs of blue smoke emitting from the tailpipe (just watch stock 5.0-liter Mustangs someday—they use low-tension oil control rings and almost always puff the blues when accelerating or starting up), because excessive oil is allowed past the rings.

As with pistons, whenever you rebuild an engine, aftermarket piston rings are the most logical choice, because of the variety of materials, sizes, cost, and even availability.

Machining and Preparation

The critical element of piston rings are the end gaps, the correct measurements for which are typically provided with the specific set of rings (or can be customized to your needs and the engine's application). An excessively large gap will allow too much cylinder pressure to blow through. Too tight a gap might result in a broken ring, if the ring expands so much as it heats that the ring ends physically press against each other. You measure the gap by squaring each ring about an inch into its respective bore (an inverted piston usually works well for this) then measure the gap with a feeler gauge. If the gap is too large, you can try swapping the large ring for one from a cylinder in which the gap is too small. If the gap is still too large, your only option may be to purchase oversize rings and file the gap to fit. To do so, you'll need a ring filer tool (a special, small grinding wheel) and remember to only grind one end of the ring (to leave the other square) and only grind a little bit at a time.

Beyond measuring and correcting ring end gaps, there's little that must be done to piston rings to prepare them for use. It should be noted, however, that the only correct way to install piston rings on pistons is with a set of piston ring expansion pliers (a satisfactory set will cost you about $10 at any Sears tool center). Likewise, you'll need some form of a ring compressor to squeeze the rings into the piston ring grooves for insertion into the cylinder bores. Several types of ring compressors are available: the band type, the pliers type, and the cone type, which is widely used by professional engine builders for its ease and minimal chance of ring breakage.

Main, Rod, and Cam Bearings

Bearings basically minimize wear between two metal parts. Oil between the bearing surface and the surface of the moving part, such as a crankshaft journal, is responsible for reducing friction. One important—and often overlooked—role of the bearings is to trap and absorb small contaminants, like bits of metal or even dirt, and prevent them from touching the crankshaft's journals, which would damage the fragile journal surface.

The bearing surfaces are relatively soft, so particles can actually embed in the surface and get stuck there, hopefully safely beneath the thin film of oil, out of reach of the crankshaft or camshaft journals.

Beyond those two functions, bearings do little more than provide the proper clearance through which oil can

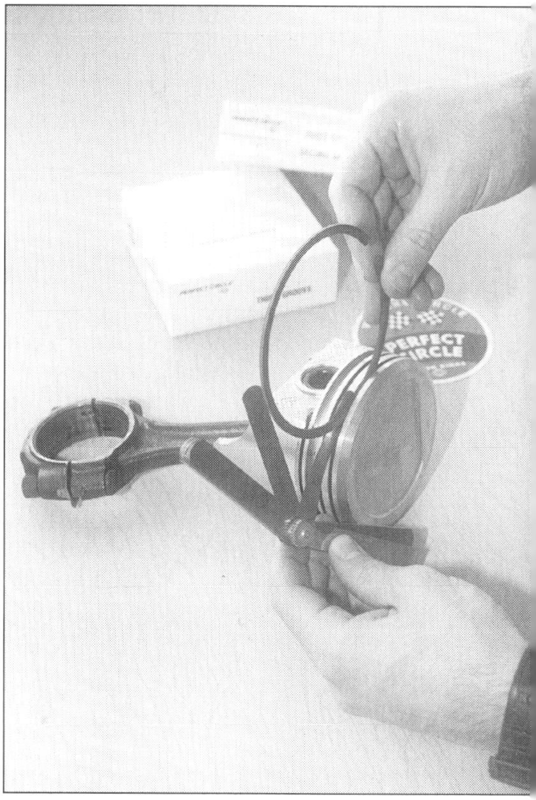

One of the first things to measure with rings is their fit within the ring grooves of the piston. Rings do come in different thicknesses, as do piston ring grooves. You need to make sure your rings and grooves are the proper matches.

Next, you need to measure the ring end gap. Start by setting the ring in the bore, like this.

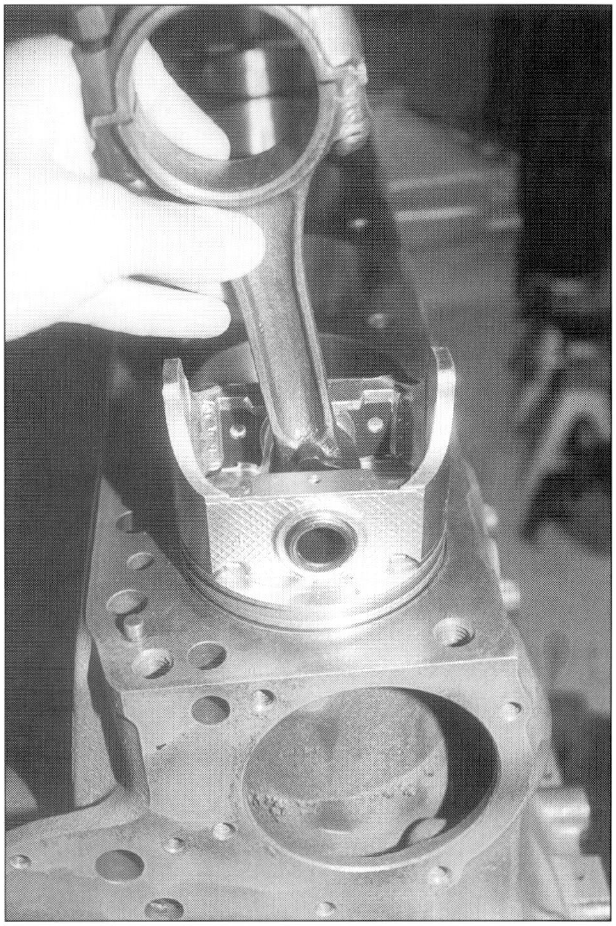

Then invert a piston and push the ring down about 1 inch into the cylinder. This squares the ring in the cylinder.

flow. But that's a critical job too. Too much clearance and oil pressure will drop, and the part (such as crankshaft) may be allowed to move excessively, leading to rapid wear. Too little clearance and oil may not be able to flow through, thus friction and wear will dramatically increase (and usually the bearing will burn up and eventually the bearing may spin in its bore or the engine may seize).

Selection

When purchasing bearings, purchase them to match any machine work that may have been done to the crankshaft. For example, if the crank was turned 0.010-inch undersize on the main and rod journals, get main and rod bearings that are designed for that change.

Bearings aren't terribly expensive, so spend the few extra dollars for quality bearings from a reputable company, such as Clevite-77.

Machining

There isn't much machine work to be performed on bearing shells. You may wish to debur the bearings *very lightly* with a ScotchBrite pad. You don't want to remove any of the bearings' surface coating, just any high spots. In addition, if the crank's journals have been radiused, the bearing edges may need to be chamfered (tapered) to avoid interference with the radiused portion of the crank journal.

It's also in your best interest to take time to match bearings to specific crank journals and block bores for the best fit. To do this, measure each crank journal with a micrometer and record the figures on a sheet of paper. Then install the bearing shells in the bearing bores and torque their respective caps in place. Then measure the inside diameter (ID) of each main and rod journal bore with the bearings installed and record those on the sheet. With your ideal clearance in mind, subtract the journal diameter from its corresponding bore's inside diameter to determine the overall gap. To optimize the measurements, you may need to try different bearing shells in different bearing bores, a process that is somewhat easier if you know the inside diameter dimension of each empty bearing bore plus the thickness of each bearing half. Thus, if you know you need a larger clearance, you could match thinner bearing shells to larger bores.

An easier method is to simply trial-assemble the engine and check clearances with PlastiGauge—thin strings of crushable plastic that you place on the journal prior to torquing the cap in place. After torquing, you remove the cap, measure

Then use a feeler gauge to measure the end gap. Compare your reading to what the ring manufacturer recommends.

the thickness of the squished plastic with the provided gauge (usually on the wrapper) to determine the clearance. If gaps aren't within tolerances, you can try swapping bearings from one journal bore to another and remeasure.

A special note on cam bearings is in order, because they differ from main and rod bearings in that they aren't easily installed: They have to be pressed into their corresponding journal bores. Given how fragile bearings are, unless you've got the proper tools for the job, you're likely to ruin a few bearings in the process.

Oil System

Most people assume your engine's oil system is only responsible for lubricating the engine, but the oil actually has five critical duties: It cleans the engine, by picking up dirt or other particles and depositing them in the oil filter; it cools the engine, just as coolant does, by absorbing heat and dissipating it to the air flowing around the oil pan; it seals the piston rings to the cylinder walls, thereby

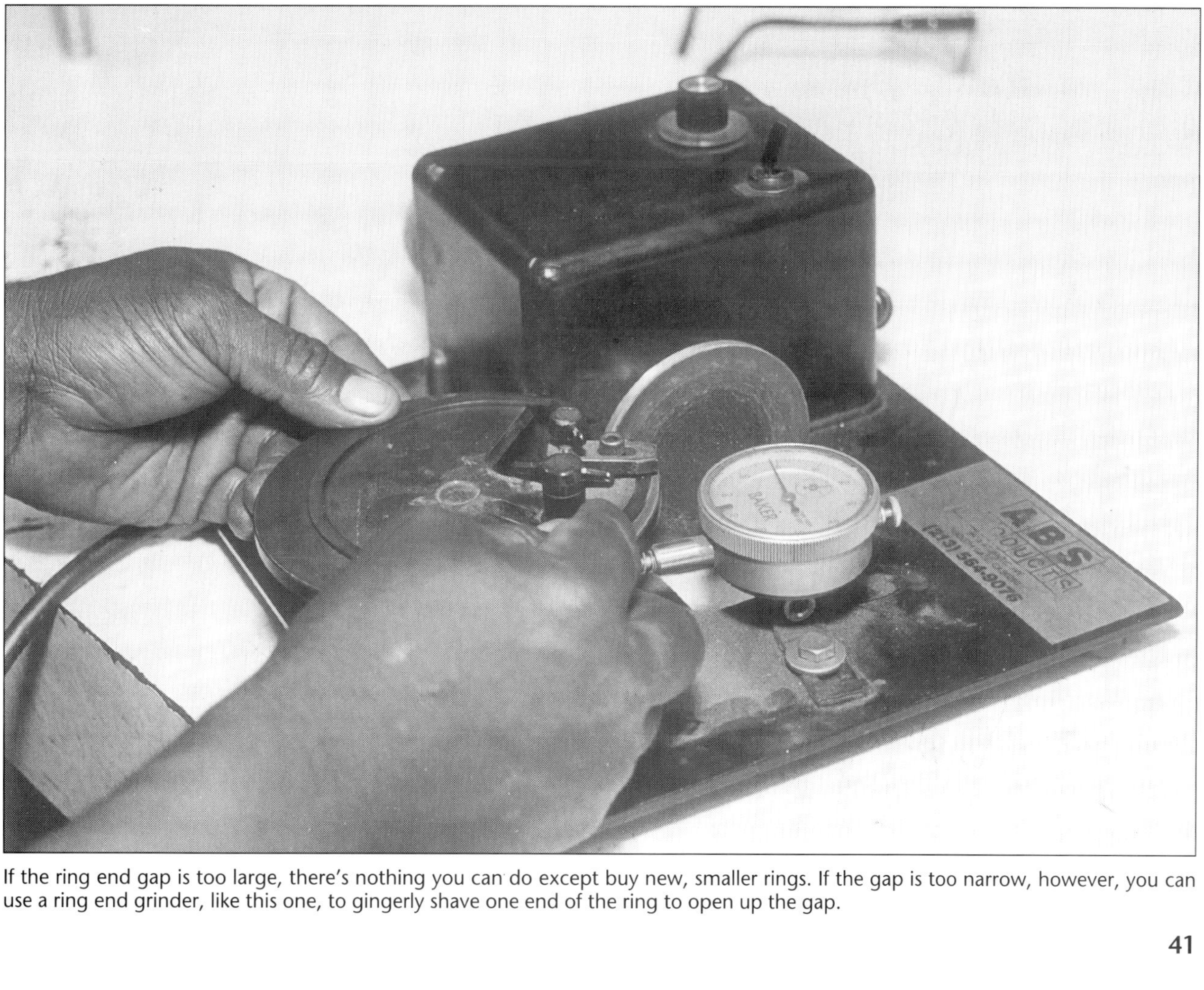

If the ring end gap is too large, there's nothing you can do except buy new, smaller rings. If the gap is too narrow, however, you can use a ring end grinder, like this one, to gingerly shave one end of the ring to open up the gap.

After installing the pistons on the rods, set a bearing shell in the rod.

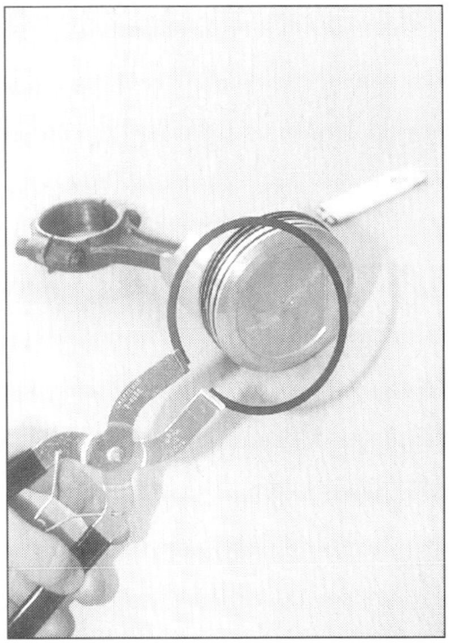

When it comes time to install the rings, ring pliers can be used to gently spread the ring to slip it down over the piston. Or you can set one end in the appropriate groove and "spiral" the rest of the ring down to the groove, but you have to be especially careful not to scratch the side of the piston.

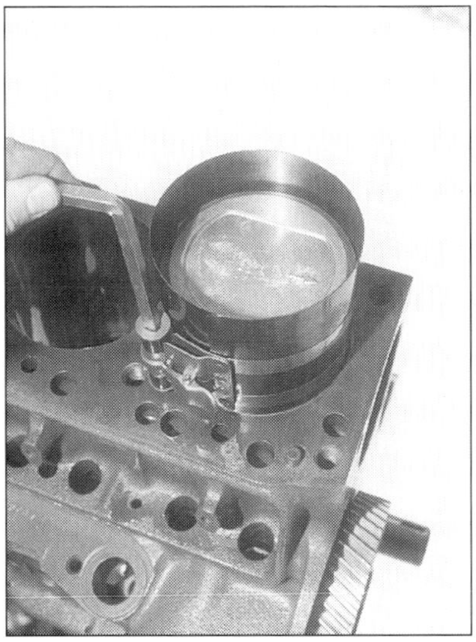

Use a piston ring compressor to tighten the rings into the ring grooves, then push the piston into the cylinder, being careful not to let the rod bolts contact the crank journal. Be sure to check each rod's oil clearance as you did with the main bearings.

helping to maximize compression; it protects metal components from corrosion and rust; and it lubricates to prevent wear between two parts. With all that to do, it becomes much more clear why a high-performance engine needs a high-performance oil system.

Oil Pan and Pickup

The most visible part of any oil system (unless you have a horrendous oil leak) is the oil pan. Stock oil pans for all small-block engines (except the LS1 and certain V-6 engines) are made of stamped steel and will typically hold 4 quarts of liquid gold, plus the filter holds another 1/2 quart. The stock pan also features a simple baffle designed to keep oil trapped in the deepest part of the pan—the sump (though the baffle does little to keep the oil pump pickup submerged in oil).

A variety of aftermarket oil pans are available that provide both increased oil volume, plus feature better baffling to prevent the oil pump from sucking air. In addition, an aluminum oil pan dissipates heat better and tends to muffle sounds better, quieting the engine.

If the particular vehicle into which you'll be installing your rebuilt engine has a low ride height, like a Camaro or Corvette, for instance, care must be taken when selecting an oil pan that the sump area won't accidentally contact the ground, which could tear open the pan, spilling oil everywhere and causing the engine to starve for oil and possibly seize—not to mention letting God-only-knows-what into your engine, like dirt, rocks, and other things that cause unspeakable engine damage.

Your choice of oil pans typically dictates the oil pump pickup and screen assembly you need to use, because the pickup's location and height are dependent upon the shape of the pan. In general, the pickup should be between 0.25 and 0.375 inch from the bottom of the pan, with its location within the sump determined by the type of driving the engine will endure. Drag cars should have the pickup at the rear of the pan because that's where the oil will go under acceleration; circle track cars will usually dictate the pickup be on the right side of the pan; street use usually means right about the middle of the pan because it has to contend with a variety of driving conditions.

Oil Pump

The stock-type oil pump draws the oil from the oil pan through a pickup tube with a coarse screen attached to the end of it to prevent large debris from being ingested. The pump then sends the oil to the oil filter, and then off to the rest of the engine through various oil passages called galleries.

The pump itself features steel gears inside a cast-iron housing. The size of the gears determines the volume of oil the pump can move in a given period of time. Oil pressure is determined by a number of factors, including a bypass spring mounted in the oil pump cover. Other factors that influence oil pressure are the oil filter and the sizes of the oil galleries within the engine. The old rule of thumb is that you want 10 pounds of oil pressure for every 1,000 rpm; so, for example, an engine running at 5,000 rpm should have 50 psi of oil pressure.

Stock-type pumps are driven by the distributor, which is, in turn, driven by a gear on the back of the camshaft. Gen-II engines don't have a traditional distributor (they use the crank-driven Opti-Spark system), but they instead feature a stub shaft that drives the oil pump off the camshaft gear. In either case, a blade on the bottom of the driveshaft mates with a slot on the pump's input shaft, and a collar couples the two together. Factory and inexpensive replacement pumps use plastic coupling collars. Better pumps use a metal collar that is pinned to the oil pump driveshaft to prevent it from breaking or falling off.

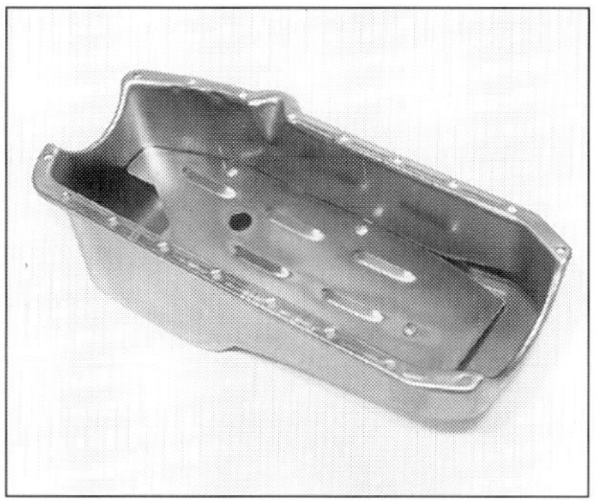

Your engine's oil system is also a part of the short-block assembly and is most easily recognized by the oil pan. This Milodon pan features a built-in baffle to help control oil in the pan. *Milodon*

Another type of oiling system, known as a dry sump system, uses a crank-driven oil pump and scavenge system. As the crankshaft spins, it turns a belt that spins the shaft of an external oil pump. One chamber of the pump actually pumps oil into the engine, while one or more scavenge chambers suck oil from a small oil pan that essentially has no sump, then pumps it to an external, remote oil reservoir tank and cooler. These ensure the system has plenty of cool oil to circulate. Dry sump systems are able to provide a more constant supply of oil to the engine, especially at high rpm, and do tend to reduce windage problems because of the scavenge stages of the pump; however, dry sump systems are considerably more complex and expensive than stock wet-sump systems, and their benefits are minimal in street-bound applications.

Windage Trays and Scrapers

Oil that isn't in the oil pan sump can't be drawn in by the oil pump, and therefore won't do the engine much good. Additionally, because oil has weight, excess oil on parts like the crankshaft can slow down the crank's rotation.

Parts like windage trays and crankshaft scrapers are designed to alleviate these problems. The windage tray, which mounts to special extended main bearing bolts, acts like a baffle to prevent oil from sloshing up onto the crankshaft as you race around, or hit the brakes. A crank scraper, on the other hand, actually skims oil off the crankshaft counterweights as it rotates.

Windage trays and crankshaft scrapers aren't used in production engines, nor are they necessary, but they do provide minute benefits if you are compelled to free up every last horsepower trapped within your engine.

Oil Coolers

Because oil flows through the engine and absorbs heat, it gets hot, and when it gets hot, it thins out, decreasing its ability to clean, seal, and especially lubricate the engine. This is especially critical for engines that will be operated for extended lengths of time, such as those used in road racing and even just street driving.

An oil cooler can considerably improve your oil's ability to do its various jobs. Factory-installed oil-to-water (actually engine coolant) coolers were available at various times throughout Camaro and Corvette production, as part of various options. The factory coolers used an adapter that mounted between the oil filter and the engine block. Coolant from the radiator flowed through the adapter, cooling the oil. It's a simple system, and it works. But it does have its problems.

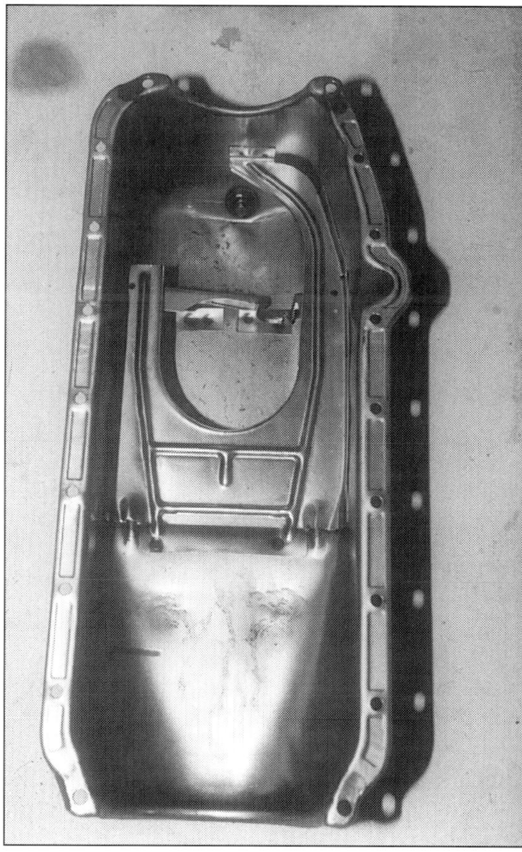

This is a factory Corvette 7-quart oil pan. Note the baffling inside the pan, which not only helps keep the oil away from the crankshaft, but also helps keep it trapped around the oil pump pickup screen.

Oil pumps are the heart of your engine's circulatory system. They push the oil from the pan throughout the engine. This is a high-volume Sealed Power replacement pump.

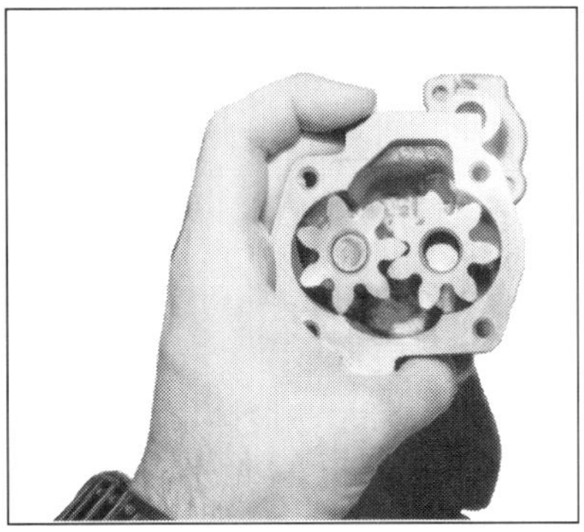

Here's a peek at the inside of the oil pump. Oil flows in to one side of the pump, then the gears, which are driven by a shaft off the rear of the camshaft, force the oil out the other side. The housing and gears should be deburred, and the clearance between the gears and the housing (including the cover) should be checked.

First, the oil is only in the adapter for a brief period of time, which means there isn't much time for the oil to dissipate its heat to the adapter, from which the heat can then be absorbed by the coolant. Second, if your engine begins to run hot, during an extended race, for example, the oil is allowed to heat up. Third, if the oil was running cool on its own, the coolant will actually heat the oil up.

Most aftermarket oil coolers are oil-to-air coolers, instead. As with the factory system, an adapter is installed between the oil filter and the block, but this adapter routes the oil out of the engine through a hose to a miniature radiator that should be mounted in an area that gets decent airflow. As with the cooling system radiator, the oil cooler radiator absorbs heat from the oil as the oil flows through tubes in the radiator. That heat is then transferred to a series of fins between the tubes, and ultimately dissipated to the air flowing through the radiator. The cooled oil then flows back to the adapter through another hose and back into the engine. Some coolers are even thermostatically controlled to prevent the oil from becoming too cool. This type of cooler system also adds considerably to the volume of oil available to an engine—often an additional two quarts or more.

Oil Filters

Oil filters are remarkably unremarkable, usually. You change your filter when you change your oil, then forget about it until the next oil change. But not all filters are created equally.

Stock filters are sealed metal canisters with a mesh-type filter material inside through which the oil must flow. As the oil flows through the mesh, dirt and metal particles are trapped in the mesh, so only clean oil flows out of the filter. Unfortunately, standard replacement oil filters do only a decent job of cleaning the oil. For one thing, the mesh can let smaller particles through that can still damage your engine. Another problem is that stock replacement filters can be quite restrictive to the flow of oil, especially as the filter traps more and more contaminants, blocking the mesh. Finally, because the stock filters are relatively small, they can quickly become plugged up with dirt and debris, allowing your oil to bypass the filter entirely if the filter becomes too much of a restriction.

Aftermarket serviceable oil filters are available that you can periodically clean and reuse. This can be advantageous because it allows you to easily see the kinds of contaminants that are in your engine's oil, and thus gives you important diagnostic information. For instance, the presence of a lot of metal particles could indicate abnormal wear in your engine. (Stock replacement-type filters can be cut open and inspected, but the process is messy and time-consuming, and the folded mesh makes inspection difficult.)

In addition, some aftermarket filters offer superior filtering properties, allowing the filter to trap much smaller particles than would a stock replacement filter.

Oil

The most important part of your engine's oil system is its oil. It is the life blood of your engine and without it your engine will come to a grinding halt in a hurry.

Crate Engines

Over-the-counter "crate" engines (so-named because they usually come shipped in large wooden crates) have been popular with performance enthusiasts since the 1950s when people would buy replacement high-performance Corvette V-8s and drop them into sedate Chevy passenger cars and wagons. In the 1960s and 1970s, Chevy created some of the hottest crate engines ever, including the big-block LS-7 454 that was rumored to produce more than 500 horsepower.

Today, General Motors Service Parts Operations (GM SPO) distributes and sells "factory" crate engines, including the latest monster, the ZZ-430—a limited edition 430-horsepower 350-cubic inch small-block that's complete from oil pan to intake manifold. All it needs is a carburetor or fuel-injection unit and exhaust manifolds or headers. It even has an ignition system right out of the crate.

The joy of a crate engine is that you can pick it up on a Friday and have it installed in the car and running the next day. You basically just bolt it in place, add oil, fuel, and electricity, and go. Crate engines give you all this convenience plus the security of a factory warranty. And since the engines are generally designed, tested, and marketed as developing a given power figure, you'll know precisely how much power the engine makes, ending the "Well, I think it makes about [X] horsepower . . ." boasts. You can state with confidence that your engine makes so much horsepower and torque at such and such rpm. What more could you want?

For starters, you might want the satisfaction of building the engine yourself, which you obviously won't get with a crate engine. You also won't know some of the exact parts used in the assembly of the engine, nor how much more power they'll necessarily withstand when you decide to make a few upgrades.

In addition, as with engine kits, the engines may not have the exact components that you want installed. So if you're determined to have those parts, you're going to have to spend extra to buy them, then at least partially dismantle the engine assembly to install them—both of which defeat two of the primary advantages of crate engines: price and convenience.

By and large, however, buying a crate engine can be a great way to boost power and performance in your car, especially if you're not as mechanically adept as you might wish to be.

It goes without saying that you need to ensure your engine always has an ample supply of clean, cool oil to properly perform, so you need to perform oil changes every 3,000 miles for street use, or after every race for competition use.

But what oil should you use? Any good-quality engine oil will generally do, though there are advantages to using synthetic oils, like Mobil 1. Synthetic oils are scientifically engineered to withstand heat extremes better—both hot and cold. They don't break down and thin out as much in extreme heat, so they protect your engine better. Plus, they flow more easily in extreme cold weather, allowing oil to flow to your engine's vital parts more quickly than could conventional oil, which would leave your engine unprotected. And tests have shown that the bulk of an engine's wear occurs during the first minute or so following start-up, so anything you can do to speed up oil delivery will dramatically reduce wear.

Synthetic oils have also been shown to result in increased power production at high rpm; however, the increase is minimal and not realizable at normal engine speeds.

As with the rear main bearing cap oil cavity, the cavity in the oil pump should be cleaned up with a grinder. Note the plastic oil pump driveshaft coupler. High-performance oil pump drive shafts use pinned-on steel couplers.

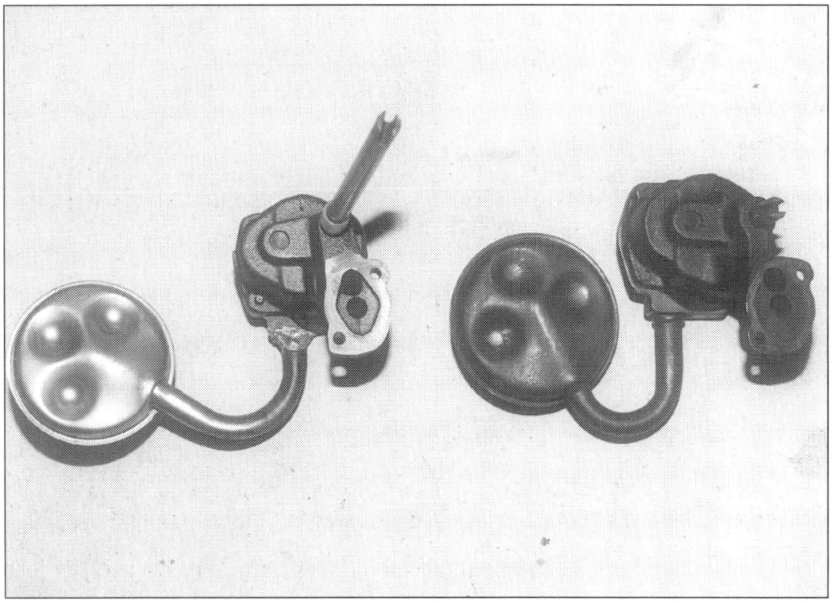

Oil pump pickups vary depending on the oil pan design. These two pumps are the same, and they're both GM pickups. The one on the left is for the 7-quart Corvette pan, and the one on the right is for a standard 4-quart passenger car pan.

Oil Additives

In recent years, a number of oil additives have hit the shelves of auto parts stores claiming to do everything from eliminating engine wear to restoring engine performance. While there is generally some truth to the claims being made, the advertisements rarely provide the whole truth behind the claims.

The bottom line on engine oil additives is that a properly maintained engine that's in good condition shouldn't need any of those additives. As proof, just look to professional motorsports teams from NASCAR, NHRA, SCCA, and IRL or CART competition: If the additives provided true, valuable benefits, you can bet those teams would be using them to gain an edge on their competition. But they don't use them (and despite what one ad claims, it's not because the rules don't allow them to). You draw your own conclusions.

Pre-Oilers

A moment ago we discussed that most engine wear occurs at start-up when the engine's oil is still down in the pan. Unfortunately, the way Chevy small-block oil systems work, you're faced with a Catch-22: Small-blocks don't pump oil until they're spinning, but you don't want to spin them until they've been pumping oil. The solution is what is known as a pre-oiler. Just prior to the initial fire-up following a rebuild (or healthy modification), it's pretty common to use an oil system primer rod (which you turn with a drill), that merely engages the oil pump driveshaft. To be truly effective, better oil pump primer rods include a collar that links two oil galleries, which are normally linked by a provision in the distributor housing, thus ensuring oil delivery to the entire engine.

But using primer rods requires removal of the distributor, and therefore isn't practical for every time you're going to start the engine. Aftermarket pre-oilers use an electric motor to pump oil from a reserve tank throughout the engine. Prior to starting the car, you hit the switch for the pre-oiler and wait 30 seconds or so for oil to circulate everywhere. Then you twist the key and your engine can fire to life relatively risk-and-friction-free because it's already protected by oil.

Selection

A higher-volume oil pan will increase how much oil your engine has available, and this can be especially important at higher engine speeds, when oil is drawn out of the pan more quickly than it returns, because it drains back slowly from the heads, lifter valley, and other areas of the engine, not under pressure. For many enthusiasts, a low-profile, high-volume pan and an appropriate pickup assembly is a good pick.

Pump-wise, a stock pump actually will suffice in many instances. If bearing clearances are slightly larger in your

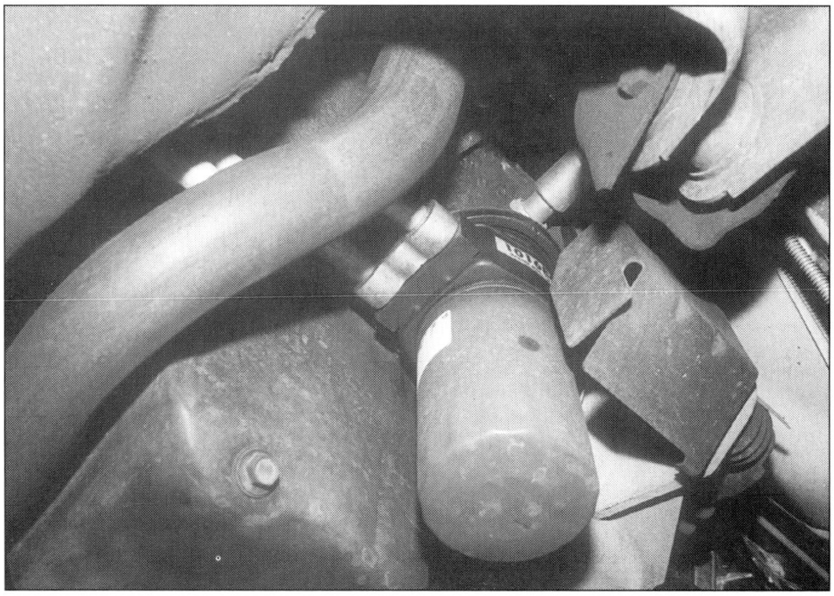

This photo shows two other important parts of your oil system: the filter and a factory oil-to-water oil cooler, which is sandwiched between the filter and the block.

Internal Versus External Balancing

To promote smooth engine operation, to extend bearing life, and to reduce harmonic vibrations that can fatigue parts, many moving engine parts are balanced to match weights of similar parts and therefore minimize opposing forces generated within the engine.

One part that gets balanced is the crankshaft, but Chevrolet has used two different methods of balancing its crankshafts over the years: adding and removing weight to the crankshaft counterweights, or adding or removing weight from the harmonic vibration damper and flywheel (or flexplate for vehicles equipped with automatic transmissions). The former method is called "internal balancing," because it deals with parts of the crankshaft that are inside the engine. The latter method is, conversely, called "external balancing" because the parts are both outside the engine and are not integral parts of the crankshaft; they're bolted to it.

Until the late-1980s, it was pretty easy to keep track of which small-block engines were internally balanced and which were externally balanced. All except the 400-cubic inch version were internally balanced, but the 400's longer stroke would have required the addition of so-called heavy metal to the crank's counterweights to compensate for the harmonics created by the long stroke.

But late-model engines such as the 305 and 350 TPI engines, including the Gen-II and Gen-III engines, also make use of external balancing. One reason for the broadened applications of external balancing is that it allows a reduction in the size of the crankshaft counterweights, which allows more compact engine designs. External balancing is also less expensive to manufacture than internal balancing, in part because it requires less costly, less time-intensive machining.

The catch, of course, is that internally balanced and externally balanced crankshafts (and their associated parts, such as vibration dampers and flywheels) are not interchangeable. Internally balanced cranks can use dampers and flywheels designed only for them; externally balanced cranks, likewise, require parts for externally balanced cranks.

The 305, in particular, has a lighter piston and rod assembly because of its reduced bore (compared to the otherwise similar 350). This necessitates a smaller counterweight to achieve a proper balance, which means the 305 crank can't be used with heavier stock 350 or larger connecting rods and pistons, unless so-called "heavy metal" is added to the counterweights to increase their mass. Note, however, that many aftermarket rods and pistons for 350s, 400s, or various "stroker" motors are as light as or lighter than 305 units, which can make a 305 entirely suitable for such applications.

engine, in anticipation of extended high-rpm use, a high-pressure or high-volume pump would make sense.

A windage tray—even just a basic one—is a sound investment, and so are oil coolers, high-performance filters, and a pre-oiler, though the latter is optional, given that countless small-blocks survive hundreds of thousands of miles without them. Oil-wise, anything will do, but synthetics make the most sense (except, perhaps, from a cost perspective).

Last, it's critical that you use a high-quality gauge to monitor oil pressure, and it wouldn't hurt to install an oil temperature gauge, as well, to alert you to any unexpected hot times. Mechanical gauges tend to react more quickly and accurately to changes in oil pressure and temperature, so they are preferable to electric gauges, though quality aftermarket electric gauges are usually more accurate than stock gauges.

Modifications and Preparation

Believe it or not, there are actually several modifications you can make to your oiling system to improve its performance and reliability. Starting with the oil pump, you can debur the pump's outlet and passages and blueprint its gear-to-cover clearance (0.002 to 0.004 inch) and gear-to-sidewall clearance (also 0.002 to 0.004 inch). Cover clearance can be corrected by milling or flat-machining the pump's mating surface the required amount to achieve the low-end of the tolerance. Adjusting sidewall clearance requires either different gears or a different case.

The oil pickup you use should be brazed or tack-welded to the pump cover to ensure it remains properly installed and doesn't vibrate loose.

The oil pan should be suitably baffled and fitted with a magnetic drain plug to help collect any metal particles in the oil, plus a crank scraper should be suitably fitted to help remove excess oil from the crankshaft.

Fasteners—Nuts and Bolts

Aftermarket bolts typically offer far better tensile and shear strength than stock bolts, allowing them to hold an engine together longer, under higher stress. Studs are even better, because—unlike bolts, which concentrate stretch forces beneath the head of the bolt—studs distribute the stretching forces throughout the length of the stud, minimizing the chance that the stud will fail.

Studs provide a secondary benefit, as well, in that they are not removed from the block during engine tear-downs and rebuilds, and therefore do not weaken the threads (by wearing minute iron particles from the threads), so the block can withstand many more rebuilds.

Recommended Modifications

By now, if you've actually read this far and didn't just skip ahead, you're probably muttering something like: Gee, this is all *really* fascinating, but what does it mean I should do with my engine? A very good question, to which there are several very good answers. Which one applies to you really depends on how you will be using your engine.

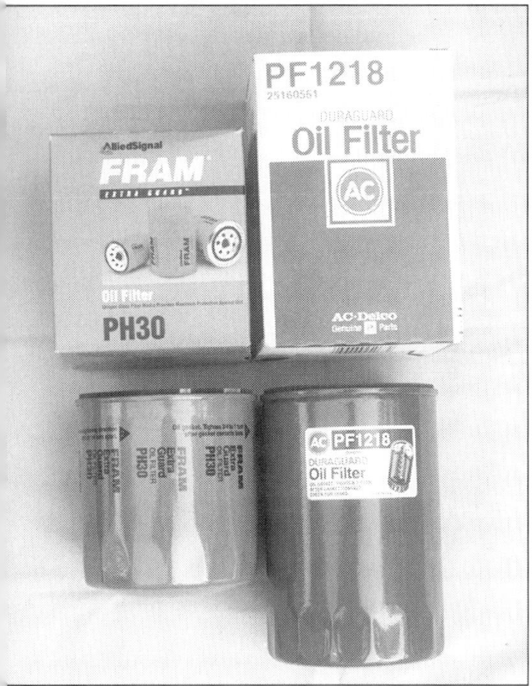

Oil filters aren't created equal. As this photo shows, there are clearly different sizes, though these two filters are interchangeable. The taller filter holds more oil and has more filter surface area for better filtration. Additionally, serviceable filters are available, as are some with oil additives applied to the filter media.

Know Your Needs

The first thing to look at is what you intend to do with your engine when you are finished with it. Which of the four basic performance applications—high-performance street driving, drag racing, autocrossing, or road racing—will the engine be subjected to most often? Obviously, you may do some weekend autocrossing or drag racing with your high-performance street car, but you need to figure out what you'll be doing with the car 95 percent of the time and build the engine to handle those situations best.

A *high-performance street* engine rarely ever develops anywhere near peak power. In fact, it spends most of its time loafing along in the lower end of the rpm scale, making enough power to scoot you along within (we hope) legal limits. The hard part of building an engine for this use is planning for the engine's survival throughout a wide variety of operating conditions: extreme temperatures (both hot and cold), light loads to heavy loads, light throttle to full-throttle, lots of short trips, stop-and-go driving, plus hours on end at highway speeds. A street engine has to be ready for just about anything, including weekend trips to the drag strip or autocross course. In this case, it pays to think first and foremost of durability: You want this engine to be rock-solid reliable and to last for many years and hundreds of thousands of miles. The key to achieving this is in selecting parts that are heavy duty and

Vibration dampeners, also (incorrectly) known as harmonic balancers, dampen vibrations caused by the natural harmonics of an operating engine. The dampener is designed to smooth out engine operation and keep the engine from shaking itself apart. This is a Gen-II small-block's dampener, which does double-duty as the lower engine pulley.

Most racing organizations prohibit the use of production vibration dampeners because they are not able to withstand high-rpm use. High-performance dampeners like Vibratech's Fluidampr are excellent for both the street and racetrack, however. *Vibratech*

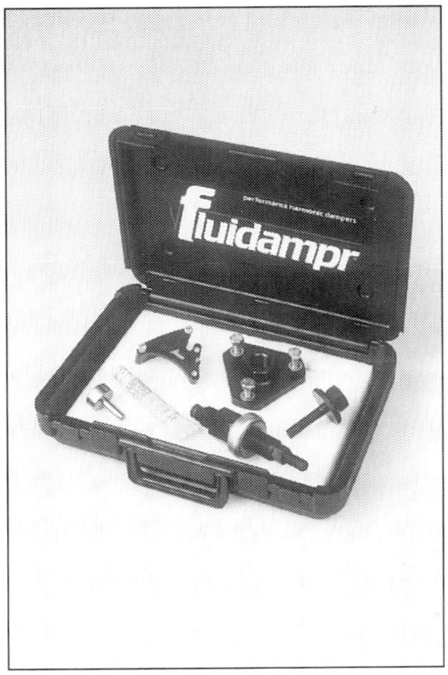

When installing and removing vibration dampeners, many people tend to use the old hammer-it-on-till-it-won't-move-any-farther method. But all that does is increase your crankshaft's end-play—probably beyond tolerance. The only right way to install and remove balancers is with the proper pullers and installers, like this kit from Vibratech, the makers of Fluidamprs. *Vibratech*

don't employ trick modifications that jeopardize engine reliability in the name of more power. The bottom end of your engine isn't the area to be making power—top-end modifications will work just fine for that.

In fact, the easiest way to prepare a short-block assembly for high-performance street use is to take the previously mentioned advice of TPI Specialties' Myron Cottrell and leave the stock short-block as is, since it can safely and reliably handle 400-plus horsepower. Even if you're working on an LT4, that allows for a 60-horsepower gain before you need to start getting too concerned.

If your power will exceed 400 horses, or if you're planning to use a supercharger or nitrous—which greatly increase cylinder pressures—you would be smart to spend a little time fortifying the bottom end.

Drag racing engines, on the other hand, only have to be able to survive for short—but extremely intense—bursts. Even if a drag engine is only rebuilt once per season, at most it will log only a handful of miles compared to a street engine. Here, the primary goal is power production with enough durability to hold together through full-throttle, high-rpm/high-load power shifts. Because of the extreme engine speeds and loads, you need a bottom end that's strong, yet lightweight, so that the engine can accelerate quickly. But since the engine only needs to last a dozen or so seconds for any given pass, it's pretty reasonable to expect lightweight parts to survive long enough to reach the lights at the big end.

Autocrossing, is, in many ways, a lot like street-driving, only intensified: You charge hard to a corner then jump out of the throttle to squirt through the corner then jump back on the gas to blast to the next turn. Throttle action is largely all or nothing: full-throttle or closed, with very little in-between. Fortunately, the runs are pretty short and the rules are fairly strict, discouraging modifications. If you choose to modify your stock bottom end (and there's nothing that says you have to), give strong consideration to parts like a stronger crankshaft, connecting rods, and pistons. Weight should be kept in mind, but not at the expense of durability. And remember that the rules typically limit the modifications that you can make that increase power (without bumping you to a higher, more competitive class).

Like high-performance street driving, *road racing* demands longevity first and foremost, because an engine that can't go the distance can never win a race. Interestingly, road racing engines and drag racing engines are forced to endure two completely different sorts of extreme driving conditions. Whereas drag racing strains an engine for all its worth during the entire run, road racing doesn't; but it does subject the engine to high-rpm for extended periods of time.

Fortunately, road racers have the ability to be easier on equipment: Instead of jumping in and out of the gas, they can roll into it; instead of full-throttle upshifts, they can allow engine rpm to decrease a little, lessening the impact on the engine and drivetrain; and as a road racer downshifts, he can "blip" the throttle to match engine rpm to coincide with the gear he (or she) is shifting into, thus preventing sudden shocks to the drivetrain induced by letting the rear wheels crank up the engine speed. All this means that, while durability is key, the engine doesn't have to be built like a tank. In fact, some lightweight parts may actually help the engine survive the extended-duration high-rpm use by decreasing momentum within the engine (you may recall from your high school physics class that momentum is defined as mass x speed: reduce either and you reduce momentum).

Know Your Budget

Next, think about your finances: How much money can you spend on your rebuild? There's nothing wrong with not having the finances to buy top-dollar aftermarket parts. The folks at Callaway Cars are well known for their abilities to take stock parts, carefully prepare them, and come away with more power in an engine than others get by bolting on all the high-performance parts they can find. It isn't always how much money you spend on parts that determines power, but rather how much time and thought you give them. Of course, depending on your driving application, it is possible that some stock parts just won't survive no matter how much time

Pulleys shouldn't be overlooked on performance engines, either. Warped or misaligned pulleys can quickly throw a belt, which on first-generation small-blocks results in a nonfunctioning water pump and an overheated engine. At the very least, you'll end up discharging your battery because the alternator won't be turning. These billet aluminum pulleys are manufactured by March Performance. *March Performance*

you spend preparing them, so aftermarket parts may be the only smart choice.

Plan for the Future

It pays to think about the fact that the short-block assembly is pretty darn difficult to remove from the car, too—especially from late-model Camaros and Corvettes, which require the engine to go in and out from beneath the cars, ruling out most do-it-yourself swaps. To minimize the chance that you'll have to pull the engine again and redo work you've already done, it often pays to set up your engine's bottom end to handle any foreseeable power upgrades. Until those upgrades come along, your engine will just be that much more reliable.

You'll notice this last philosophy in the following recommendations: There are relatively few suggested differences between a high-performance street setup and a road race setup. The recommendations are for a sort of one-size-fits-all short-block assembly that's strong enough to survive any application. Differences come into play for two primary reasons: budget and durability. Some race parts are just outrageously expensive, and thus don't make sense for a street engine, so you'll find practical street alternatives provided where they make sense. Durability-wise, however, race engines often require frequent rebuilds, which aren't exactly practical for street use, so things like clearances and even some parts choices should be chosen with the appropriate application in mind.

As you read through the remainder of this chapter, remember that your particular needs—and your finances—may dictate or allow variations from the following recommended setups. And that's okay, as long as you think things through.

Prepping the Parts

Block

While a two-bolt block will suffice for street use, a four-bolt block should be considered preferable and also more than enough for the street. Four-bolt blocks are a virtual necessity for racing purposes, and if you'll be competing for money, it might be wise to consider a splayed-bolt four-bolt main cap arrangement for added strength and rigidity. In every case, thoroughly debur the block, lightly round every edge (except around the bores at the deck surface), remove excess casting flash, and smooth and/or paint the lifter valley to promote oil drainage. Only overbore the block if needed, because the process removes material, weakening the block. Make sure the block is dimensionally correct (main journal diameters and alignment, deck flatness, bores properly sized and round top to bottom, and so on). It's also smart to replace stock fasteners—main bolts, rod bolts, and head bolts, specifically—with high-performance fasteners. Main and head studs are actually preferable, but bolts will work fine in all but the most strenuous cases.

Crankshaft

Herb Couzins at Callaway Cars strongly advocates investing in a quality crankshaft, saying that some things just aren't worth skimping on. A high-performance crank, like those from Callies, Lunati, or several other manufacturers, is like valuable insurance—you'll never have to worry about it failing. But if your budget can't handle $1,000 for a crankshaft, you can always modify the one you've got (or a less expensive aftermarket unit) to increase its performance capabilities. Enlarging the journal radii, deburring it, balancing it, and hardening it will greatly increase its strength—and your peace of mind. Still, if you do splurge for an aftermarket crank, it makes sense to get a slightly longer stroke piece, since there is little if any difference in price, but substantial power gains to be had. Note, however, that longer-stroke cranks do require pistons with a correspondingly shorter piston pin-to-deck height, to avoid having the pistons pop up out of the cylinder bore.

A purpose-built high-performance crankshaft should be considered a must for race-only applications. They're stronger, lighter, and usually have the bulk of the machine work done for you, which at least saves you some time. Be sure to check any applicable racing rules, however, before considering a stroke change that would affect displacement.

Connecting Rods

Check with your chosen machine shop before you decide what to do about connecting rods. To properly prep them, your rods will need to be Magnafluxed, be resized, have new bolts installed, have the area near the bolt head radiused, be completely deburred, be balanced, and then be fitted to the pistons and pins. Find out what all that work will cost you, then compare that to the cost of buying and mounting new rods that have most of that work already done. If there's little difference in price, go for the aftermarket rods, which will usually prove to be stronger and lighter weight. Note, again, that if you're buying new rods anyway, you may want to consider getting slightly longer rods, which reduce rod angularity for reduced friction and increase piston dwell for a more complete burn (longer rods require shorter pistons, though).

Certain fasteners, because they're so hard to access once an engine is installed, deserve some special attention to ensure they won't loosen over time. Stage 8 makes these nifty locking fastener systems for a number of uses, such as flywheel bolts. The bolt is tightened, then the egg-shaped lock is installed over the head, then a C-clip holds the lock in place. *Stage 8*

Pistons, Wrist Pins, and Rings

New pistons are almost a given in a rebuild, as are new wrist pins and rings. While hypereutectic pistons are quite good—especially if you'll be using your engine in a daily driver—forged pistons are stronger and comparably priced. A number of factors will dictate the pistons you select, including bore size, stroke length, and rod length. You will need to consider your desired compression ratio (and thus combustion chamber size, too), but in general flat-top pistons are best for making power, and you should keep compression under 10.0:1 for any vehicles that will run pump gas, while racing class rules will usually dictate the maximum compression ratio allowable for your chosen field of competition.

Wrist pins usually come with the pistons and should work well, but if your engine is a race motor, you might want to consider light-weight pins. Bear in mind that light-weight pins have less beef to them, so they're not as strong as a heavier pin of the same material. Naturally, the pins you use have to be the appropriate design to match your pistons and rods: full-floating, floating in the rod, or floating in the piston.

As for rings, plasma-coated rings of the appropriate size don't require much run-in, they seal well, they're durable, and they're reasonably priced. And unless you're building a race-only motor, it's easier to stay away from file-to-fit rings, unless, of course, you need them to achieve a proper gap.

Oil System

A smart setup for your oiling system is to run an oversize pan that adds a quart or two capacity, to help prevent oil starvation during performance driving situations. A high-volume oil pump with an appropriate pickup, a windage tray, and an oil cooler should round-out the hardware. Make sure you debur everything, and clean up oil passages on the rear main cap and inside the pump itself for better flow. Painting the nonmachined interior surfaces of the block can aid oil drain back, too, though it isn't absolutely necessary. When it comes to oil, it's largely your call. Synthetic oil does provide slightly better protection under hard use, but if you change your oil every 3,000 miles, conventional oil works perfectly well. Finally, always use a quality oil filter, and don't hesitate to change it more often than you change your oil.

Main, Rod, and Cam Bearings

Given that it only makes sense to choose quality bearings from a reputable manufacturer, and that you're ordering them in the proper size, your only real concern is to make sure the clearances between the bearings and their respective journals (main, rod, or cam) are within acceptable limits.

Wider clearances will reduce friction (to a point), but will eventually pound out the bearings due to increased movement of the part. Narrower clearances tend to tighten up under use, which reduces the flow of oil between the bearing and part, increasing friction. As long as the clearance isn't too tight, however, wear is minimized because the part can't move around much—a constant, pressurized oil film keeps the part suspended off the bearing surfaces.

For street use, keep the clearances toward the tighter end of the accepted ranges to increase engine durability.

Racing applications—including drag racing, autocrossing, and road racing—will benefit from clearances toward the looser (larger) end of the recommendations.

CONTENTS

CHAPTER THREE

CYLINDER HEADS

At their most basic level, engines are merely air pumps—they draw air in, mix a little fuel with it, burn it, then blow out what's left over. That being the case, the more you can do to improve your engine's aspiration abilities, the better it will perform. That's why devices like superchargers and turbochargers work so well—they super-aspirate the engine, and it produces super power in return.

Understanding Cylinder Heads

While a number of engine components play significant factors in determining your engine's overall breathing abilities, none are more critical than the cylinder heads. The primary reason the heads are so all-important is that they are the last possible restriction for air going into the engine, and the first possible restriction for exhaust going out. As such, no matter how free-breathing your induction and exhaust systems are, their effects will be nullified by poor heads.

Another way of thinking about cylinder heads is to consider that ideal cylinder heads (if such things existed) would allow an engine to produce its maximum power. Thus, good cylinder heads don't really *make* more power, but rather bad heads *restrict* power; the good heads are just a lot closer to being ideal heads, so they don't limit horsepower as much as the bad heads.

Of course, this is a gross oversimplification, since a good head for street use would be a horrible head for professional road racing competition. Like most other engine components, cylinder heads have to be selected based upon the application for which they will be used.

Five aspects of cylinder heads determine how well they will contribute to power production: intake ports, valves, combustion chambers, exhaust ports, and the material from which the head is made.

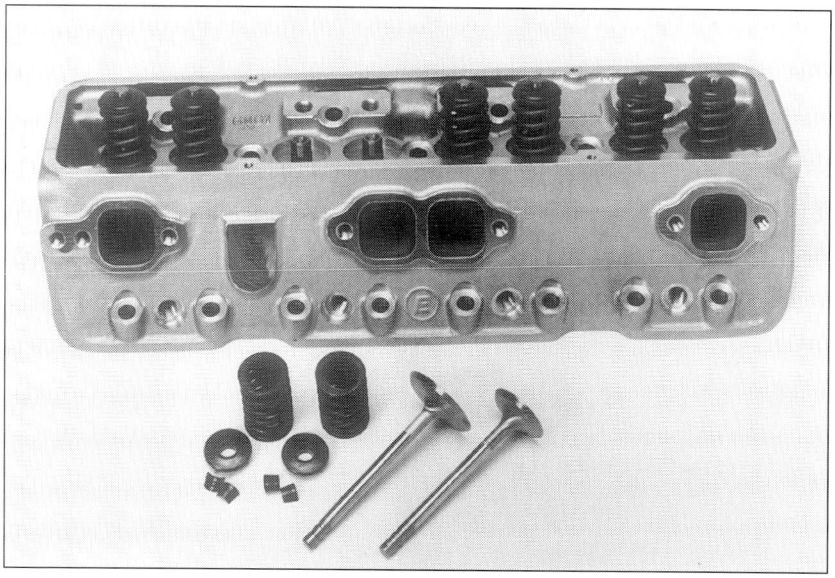

Cylinder heads ultimately determine how much air can get into and out of your engine, so they ultimately determine how much power your engine can produce. This is one of Edelbrock's aluminum Performer heads, which comes complete with valves, springs, retainers, and keepers.

Fully assembled with roller rockers and CNC-ported, Callaway Cars' aluminum SuperNatural cylinder heads are the keys to the engine's 404-plus horsepower.

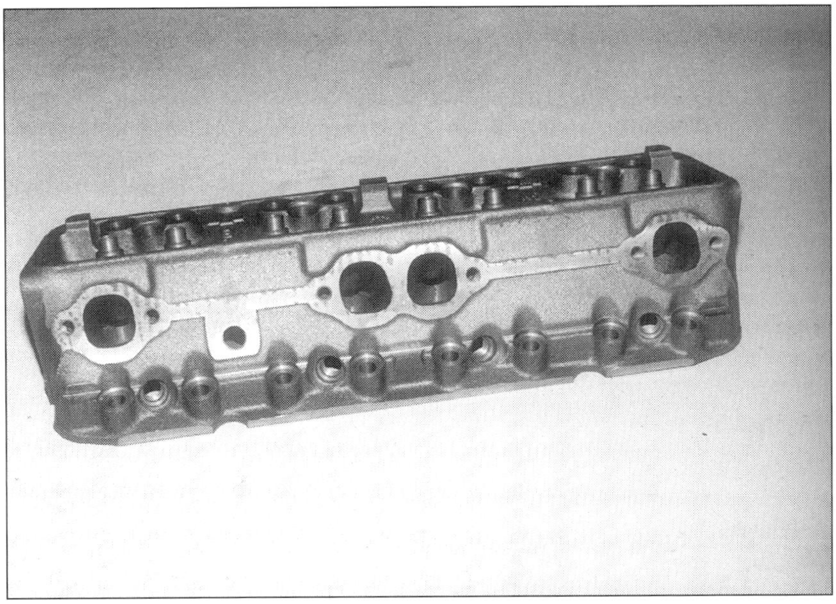

The bulk (and heft) of cylinder heads for small-block Chevys are made of cast iron, like this one.

Intake and Exhaust Ports

Intake ports receive the air from the intake manifold runners and guide it toward the cylinder, while exhaust ports take the exhaust gases from the cylinder and shuttle them to the exhaust system. How effective the ports are at doing their jobs depends on their overall shape and size. The goal is to minimize restrictions, which sounds an awful lot like "make them bigger," until you stop to think about what actually goes on inside your engine and how port sizes affect airflow.

Quite simply, small-ports flow less air, but at higher velocities than large ports. Conversely, large-volume ports have the ability to flow lots of air, but the air moves through the port more slowly.

Here's a metaphor that might help this make sense: Picture an old-fashioned bucket brigade where people are in a line passing buckets of water in order to fight a fire. If all the buckets hold 1 gallon, everyone can pass the buckets quickly, but none dump much water on the fire.

Aluminum cylinder heads, however, offer numerous advantages, not the least of which are better heat dissipation and greatly reduced weight.

With the exception of the Vortec 6000, all Gen-III small-block heads are aluminum and of a radically different design than either first- or second-generation small-block heads.

So someone decides to switch to bigger, 5-gallon buckets, figuring the fire will get put out faster that way. Unfortunately, the people on the brigade line can't handle the 5-gallon buckets as easily, so the line moves more slowly as people struggle to pass the bigger buckets. In the end, using the smaller buckets got more of the fire put out in a given period of time than the bigger buckets.

That's a lot like the way cylinder head ports work at low- and midrange engine speeds: Smaller ports may not be able to flow the same volume of air in a given period of time (and at a given pressure) as heads with larger port volumes, but smaller ports provide a ramming effect that increases cylinder filling at those engine speeds by virtue of higher airflow rates. Assuming the brigade line could pass either size bucket at a high rate of speed, however, the brigade would be able to put out more of the fire (i.e., make more power) with the larger buckets in the same amount of time than with the small buckets.

So at lower engine speeds, smaller ports are more efficient, while at higher rpm the small ports can't move enough air through the engine, so power drops off.

Though it doesn't take port shape into account, one of the most common and effective ways of determining whether a particular cylinder head will work well for a given application is to examine the head's intake port volume.

In general, three flow ranges are typically used to group small-block Chevy heads: ports of 179 ccs or less are considered small-port heads and typically work very well in street engines. Medium port volumes range from 180 to 199 ccs and are better for larger-displacement street small-blocks or smaller engines that may need to turn high rpm somewhat frequently. Finally, port volumes of 200 ccs or more are classified as large-port heads and are usually race-only pieces.

Nearly all production heads fall into the small-port category—often far below the 179-cc cut-off. The aluminum L98 Corvette head used from the mid-1980s through 1991 is a prime example with its 163-cc intake ports. Many performance heads fall into the second grouping, including GM's own LT4 head with its sizable 195-cc intake ports for improved flow. Heads in the 200-plus category are often racing-style heads, though, interestingly enough, the unique tall and skinny cathedral-style ports of the Gen-III LS1 heads land them in this group by virtue of their 204-cc intake ports.

In actuality, there is a fourth group—the *true* racing heads—that has massive intake port volumes. Good examples of these heads are GM's 18-degree and SB2 NASCAR heads, the former of which features 258-cc intake ports in CNC form, while the latter can stuff a full 280 ccs of air and fuel in its intake ports (in CNC configuration). These heads have such cavernous ports that using them on the street would simply be foolish, unless you were to start with non-CNC versions, which measure 210 and 196 ccs of intake port volume each, respectively, then do only minimal machine work.

Port shape can have just as much effect on airflow through the port as port volume, however. Ultimately, ports should feature only gentle bends that guide the air to or from the cylinder. If the airflow encounters an abrupt, tight corner, it simply smashes into it, disrupting the flow. But a smooth, gradual corner bends the air without disrupting flow.

Unfortunately, gradual corners require more room than tight ones, and the physical size of stock-style cylinder heads imposes certain unavoidable limitations. Chevrolet's first attempts at straightening out the small-block's cylinder heads involved raising the port entry (intake) and exit (exhaust), resulting in the so-called raised port cylinder heads. The improvements were promising, so Chevy went a step further and developed the high-port race-only heads, which had ports elevated roughly 1/2 inch above their stock location. In order to make such a radical change, however, the heads are much taller than production heads.

Another aspect of port shape—one that Chevrolet exploited in late-model heads—is that ports do not necessarily have to be a uniform size from top to bottom. Chevy cinched in the top of the intake ports on LT1 heads, which creates a sort of venturi-like effect that helps increase the air velocity through the port, aiding throttle response and low-rpm flow. Because the heads have a broad floor area, however, they can still flow relatively well at mid- to upper-rpm.

The design worked so well on the LT1 that Chevy also employed it on the 1996–1998 Vortec V-8s, and in the LT4

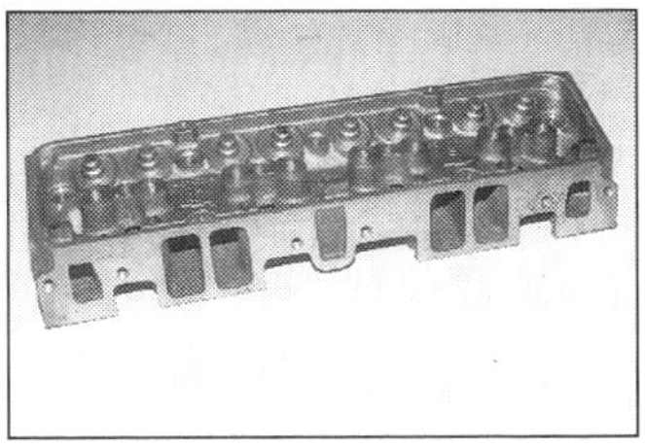

Going in order of how your engine's heads deal with things, we'll start with the intake ports. On first-generation engines, the intake ports were basically rectangular in shape. Also note the small holes at each end of the head, which funnel coolant into the intake manifold. In the center of the head, most intakes have a heat crossover passage; however, this head does not.

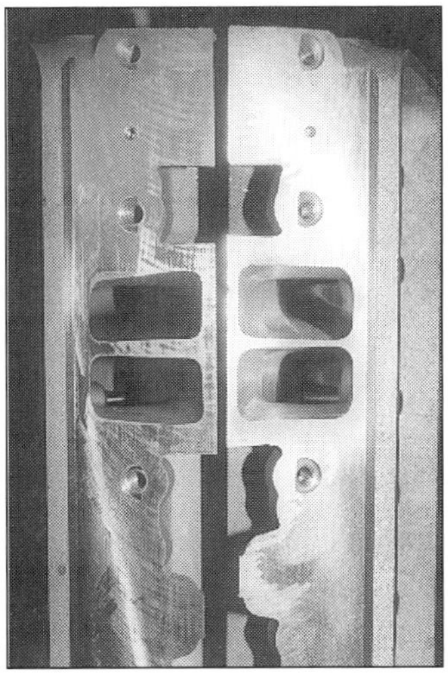

Stock Gen-II and pre-1999 Vortec V-8 ports have a trapezoidal shape, which is narrower at the top than at the bottom to higher flow velocities. Note how much larger the CNC-ported Callaway Cars SuperNatural head's ports on the right are, yet they still retain a slight trapezoidal shape. Also note the lack of the coolant passages—a dead giveaway of a Gen-II reverse-cooled head.

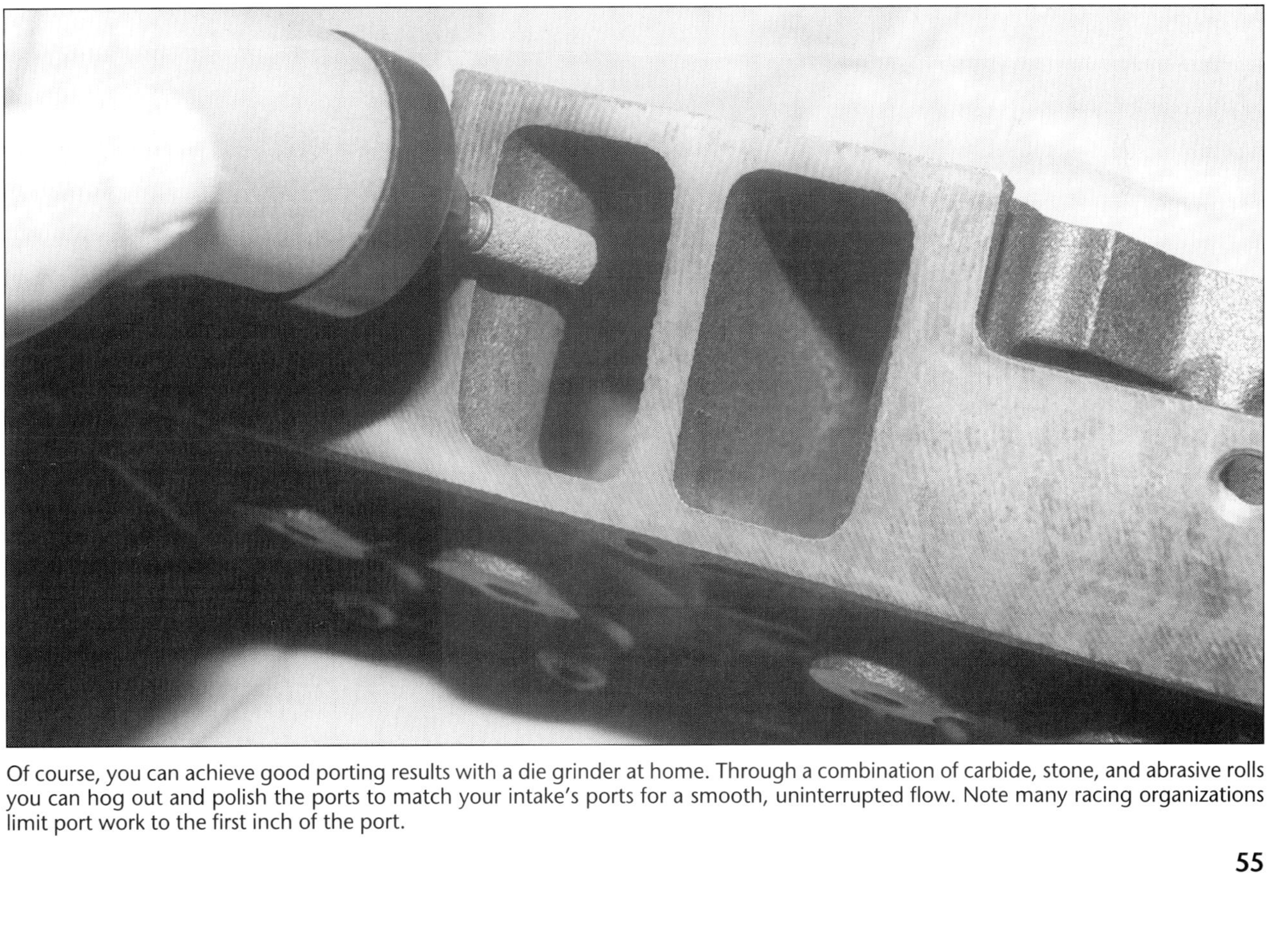

Of course, you can achieve good porting results with a die grinder at home. Through a combination of carbide, stone, and abrasive rolls you can hog out and polish the ports to match your intake's ports for a smooth, uninterrupted flow. Note many racing organizations limit port work to the first inch of the port.

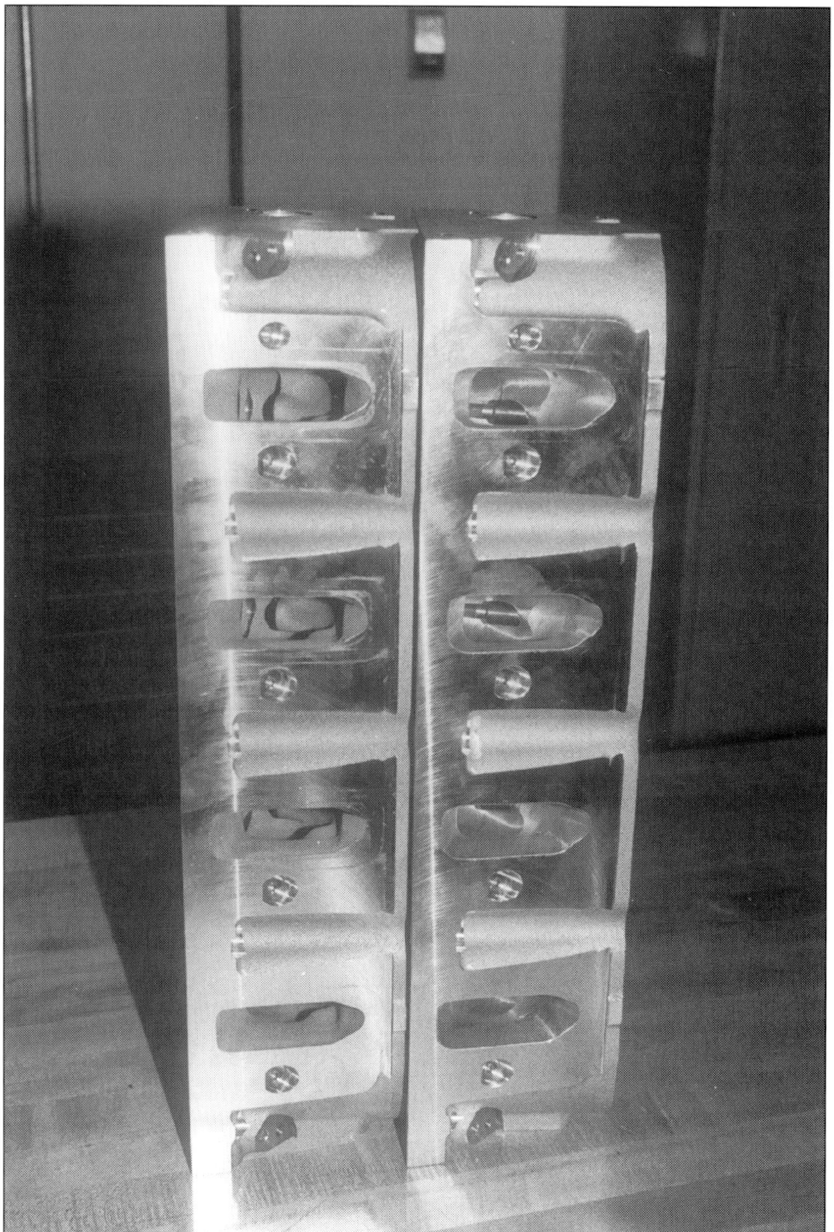

Gen-III heads have dramatically different intake ports. Referred to as "cathedral"-shaped, the ports are tall and skinny with a slight hump at the top that corresponds with the fuel injector's spray path. A stock head is on the left, while a Callaway Cars CNC-ported head is on the right. Note the huge difference in port size; an engine needs to flow a lot of air to make roughly 450 horsepower, as the Callaway SuperNatural LS1 can.

From the intake ports, the next major area is the combustion chamber, where the air and fuel are actually burned. Note the generous 64-cc size of this head's chamber.

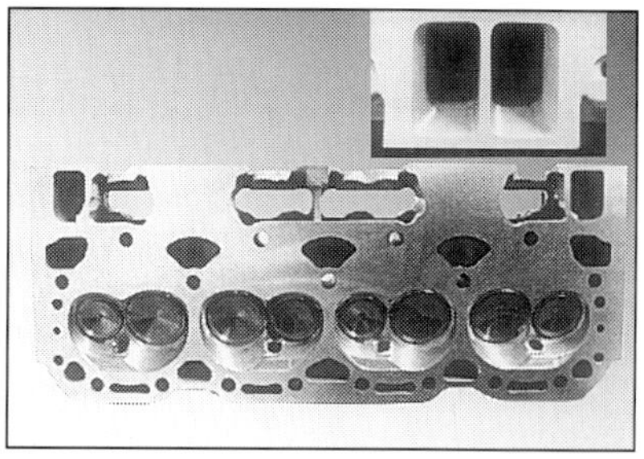

Newer cylinder heads, like this Air Flow Research aluminum head, typically feature smaller combustion chambers, which work in harmony with flat top pistons to maintain decent compression but provide an improved combustion process. Note the kidney bean shape of the combustion chamber. *Air Flow Research*

heads, though the LT4 ports were also enlarged somewhat for more volume and thus better upper-rpm flow.

An area that has traditionally presented a significant restriction to flow through small-block ports is the area known as the valve pocket—the area just beneath the valves in the ports. Quite simply, the valve pockets on production heads are too small and restrictive. Unfortunately, this is one of the more difficult areas to spend time modifying, so most enthusiasts fail to do any "pocket porting," when, in fact, small changes here can be more beneficial than large port entry and gasket-matching changes. One basic head porting guideline is to try to enlarge the diameter of the pocket area beneath the valve to between 70 and 85 percent of the valve's diameter. For a typical 2.02-inch intake valve, that means a pocket diameter of approximately 1.6 inches, while the customary 1.60-inch exhaust valves should work nicely over a 1.20-inch pocket.

A second general recommendation is to increase the short-turn radius—the area of the port floor where the floor curves downward toward the valve—by carefully removing material. A larger short-turn radius creates a straighter airflow path to the valve (often referred to as "line of sight"), which creates less turbulence beneath the valve, in the pocket area.

The valve guide support material in the valve pocket also tends to impede airflow, so streamlining the support will reduce the restriction, improving flow and ultimately increasing power.

Polishing ports until they have a smooth finish is not actually as necessary as it was once believed to be. While it is still beneficial to remove any extraneous

Head-Swapping Caveats

During the first 30 years or so of the small-block Chevrolet engine's life, one of its greatest strengths was the ability to use a part from one small-block on another. Until the mid-1980s, about the only limiting factors were some pistons, because of different bore diameters, and certain crankshafts because of changes to journal diameters.

But as emissions requirements and economic pressures lead to advances in engine engineering, the small-block Chevy benefited from improvements. Unfortunately, some of these improvements eroded the engine family's legendary interchangeability, specifically with regards to cylinder heads. This means that it's a bit more risky these days to just assume that a set of small-block heads will fit your engine. They may not.

One of the biggest differences for first-generation cylinder heads—one that actually came about in 1970 when Chevy introduced the 400-cubic inch small-block—was that the block and heads are designed with steam holes between the siamesed cylinder bores. Small-displacement small-blocks don't require these steam holes because their water jackets go completely around the cylinders. Small-displacement heads can be modified to include the steam holes; simply use a 400 head gasket to properly locate where they need to be and drill them.

When the 3.875-inch bore, 305-cubic inch V-8 came along, engineers had to modify the small-block head slightly by moving the valve stem centerlines slightly closer together. The move was primarily to eliminate the chance of a valve contacting the block deck or cylinder wall surfaces. Thus, you cannot simply bolt 350 heads onto a 305. You can, however, install 305 heads on a 350—but there's little advantage to doing so since the closer valve arrangement limits the maximum size of the valves.

A subtle change came next: the switch to "center-bolt" rocker covers from the old-style "perimeter-bolt" covers. Thanks to the new heads' redesigned gasket surfaces and center-bolt pattern, valve cover oil leaks became a rare exception rather than a common occurrence.

The next big change came in 1992 when the LT1's reverse-cooling system eliminated the cylinder head coolant passage outlets that had previously fed coolant to the intake manifold coolant crossover. This difference precludes first-generation heads from being used on Gen-II blocks and vice versa.

Last are the Gen-III cylinder heads, which are of a completely new design that is completely incompatible with first- and second-generation small-blocks: The head bolt patterns are different; the intake port locations and shapes are completely different; the exhaust ports are in different locations; the valvetrain geometry is different; the coolant passages are different. Gen-III heads will only fit Gen-III blocks.

casting irregularities (e.g., casting flash, casting marks, and so on), minor port surface irregularities induce little turbulence and thus only negligibly impede airflow through the port. And on throttle-body injected engines—as with carbureted engines—a glassy-smooth port finish may actually make it easier for fuel to fall out of suspension, thus decreasing efficiency.

Valves

All production Chevy EFI engines (except the ZR-1 Corvette's LT5 DOHC engine) use two valves per cylinder—one intake and one exhaust—to control the flow of air and fuel into the cylinder and the flow of exhaust gases out of the cylinder. Stock valves are typically steel and the valve face diameters vary depending on the particular engine.

Stock valves are quite reliable (at modest engine speeds) but often limit performance because of their smallish sizes and their inefficient valve seat and face cuts. While a performance valve job will increase the efficiency of stock valves, the size of the valve may still pre-

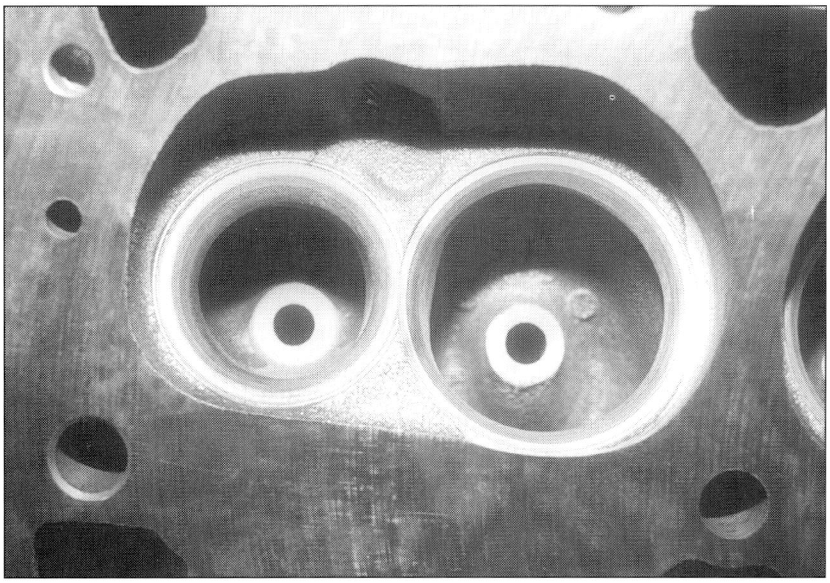

Another area to work on is the valve bowl area, the area beneath the valve in the ports. Note how the diameter of the bowl is considerably less than the diameter of the exhaust valve. Generally, if you can open this up to 70 to 80 percent of the valve's diameter, you will greatly improve flow.

The shiny areas in this chamber's bowls were opened up with a grinder.

sent a restriction to airflow. In addition, stock valves are not designed to withstand higher cylinder temperatures associated with high-performance engines.

Aftermarket valves are available in a variety of materials, the most common of which is stainless steel. They're also available in a range of sizes. The particular size of valve you run technically depends upon the design and airflow capabilities of your cylinder heads, as well as your camshaft specifications. In addition, on small-bore engines, such as the 305-cubic inch small-blocks, large diameter valves may actually contact the cylinder walls because of the close proximity of the valve's centerline to the cylinder wall. Naturally, contact between the valve and the cylinder wall would cause immediate catastrophic engine damage.

For most performance-built Gen-I and Gen-II 350-cubic inch small-blocks, 2.02-inch intake valves are desirable, while 1.60-inch exhaust valves will provide excellent cylinder emptying capabilities. Larger displacement stroker engines may respond well to larger valve sizes, though the small-diameter valves—including stock-sized 1.94/1.50-inch pieces—may serve to keep airflow velocity higher than would a larger valve, so actual airflow (volume) through the valve may decrease with larger valves.

In addition, your particular cylinder heads may preclude the use of large-diameter valves. For instance, the production aluminum L98 Corvette head won't accept any valve larger than 2.00 inches on the intake side and 1.56 inches on the exhaust side, unless you go to the trouble of having the factory-installed seat inserts replaced. Consult your cylinder head manufacturer or machinist for their recommendations.

Both stock and aftermarket valves respond well to a multi-angle valve job that smooths the flow of air into and out of the cylinders. A triple-cut valve seat and double-cut valve face typically work best, but only when performed properly. A poor multi-angle valve job can decrease flow, as well as shorten the life of your valves and seats.

Valve jobs are more science than mysterious art form, so it pays to have your heads worked on by someone with considerable experience working on performance engines. For a cheap rebuild your local machinist is probably fine, but if you're hoping to make lots of power, a valve job is not a place to cut corners, if you'll pardon the expression. A poorly performed valve job can kill a head's flow, plus lead to valve sealing problems and even durability problems.

A number of exotic valve designs and materials are available, as well. While some designs, such as hollow-stem valves that reduce weight, offer obvious advantages, they are not always well suited to street use. Necked-down valves have a smaller stem diameter for the portion of the stem that runs through the port. Necked-down valves do save a minimal amount of weight, but their true benefit is reduced flow disruption. And since they are generally durable, they make an excellent pick for nearly any application, with the possible exception of maximum effort competition engines.

Valve Guides

Valve guides are cylinders through which the valves travel. Production guides are

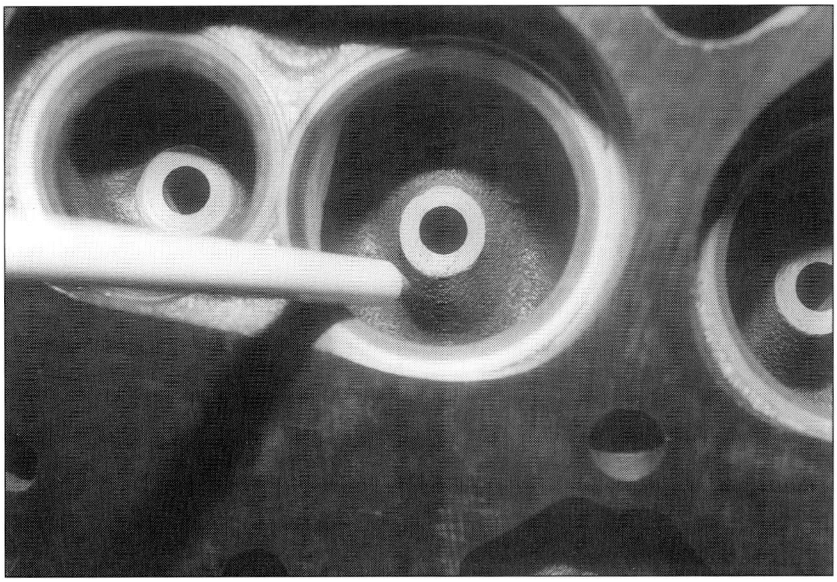

While you're working in the valve bowl, you should try to reduce the size of the valve guide support area. The supporting material is much larger than it needs to be (most CNC-ported heads don't have any supporting material left) and poses a significant restriction to flow.

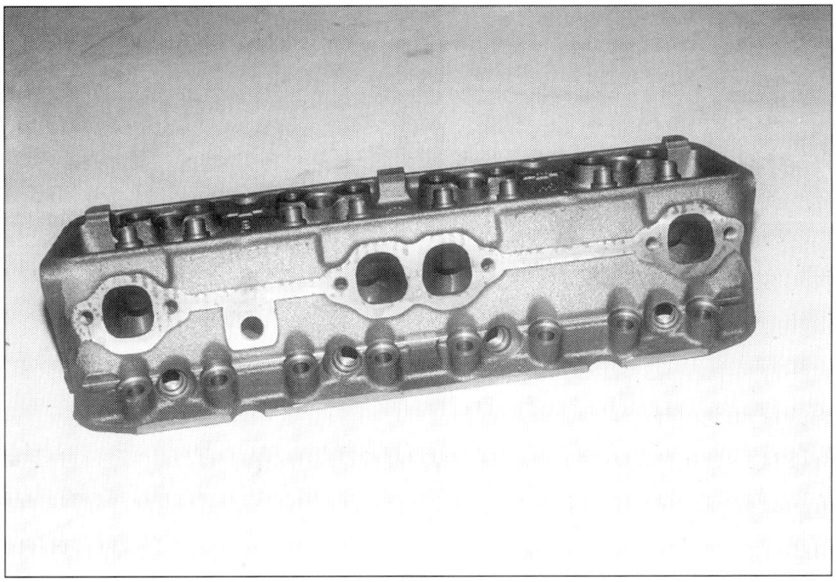

Exhaust ports in small-block Chevy heads tend to be fairly restrictive. They're smaller than intake ports because not as much air has to move through them (and the air that does is pushed out by the piston on the exhaust stroke). Note the fairly round shape of this head's ports.

iron (integral in cast-iron heads), but bronze valve guides are popular replacements for performance engines. For iron heads, many machine shops prefer to ream out the iron guides and fit them with bronze liners; the iron shell provides superior strength while the bronze lining reduces valve wear and friction.

Guide liners are an excellent solution, and one of the best designs is that of the interrupted spiral guide, which helps increase guide oiling for decreased friction and valve wear. A groove spirals down the inside of the guide liner, which allows oil to keep the valve stem well lubricated. The groove, however, does

Go with the Flow?

While the bulk of this book's coverage of cylinder heads deals with the issue of intake port volumes, the reality is that it is flow you are after, hence the coverage of porting and port shapes.

When comparing aftermarket cylinder heads, it can be quite enlightening to compare flowbench tests, which spell out airflow velocities through a head's ports. Note that you need to compare apples to apples, so make sure the tests are comparing air velocities at the same valve lift intervals. The goal is actually to find a head that will flow well at below the peak valve lift your cam and valvetrain will deliver. Don't just look at the flow figure at your cam's peak lift because the valve will only be at that point briefly, and only once. If your cam delivers a 0.500-inch valve lift and you compare figures at 0.400 inch (a fairly common figure to use for street engines), you know the valve will be at 0.400 inch twice—once while opening and again while closing.

Heads that have been ported can be tested on a flowbench as well, though be aware that your local machine shop isn't likely to have one.

Another worthwhile head comparison figure is the E/I ratio, which expresses the percentage of exhaust port flow compared to intake port flow at the same valve lift. For example, an intake port that flows 200 cubic feet per minute at a 0.400-inch lift and the exhaust port flows 150 cubic feet per minute at a 0.400-inch lift, the E/I ratio is 75 percent, which is quite good. Many production and aftermarket performance heads are between 60 and 70 percent E/I, though some aftermarket heads do have an E/I relationship of 80 percent or more.

Raw, Rough, Bare, or Assembled?

As if buying high-performance cylinder heads wasn't complicated enough with all the different port volumes, port heights, chamber volumes, chamber shapes, spark plug angles, and other variables, you're also likely to come across heads that are available in varying stages of readiness, from raw castings to rough-machined pieces, to fully machined but bare units to fully assembled heads that are ready to bolt on and run.

Raw castings are, as their name implies, unfinished pieces fresh from the casting process with no machining performed. Ports are of minimal size and only roughly shaped. Chambers, likewise, are only partially formed and typically won't have a hole drilled for the spark plug. Furthermore, spring seats, rocker stud mounting bosses, and other provisions are all left for you and your machinist to work out.

Rough-machined heads are a step above raw castings in that they feature some machining—chamber shapes, but not valve seats, for instance—to minimize the expense and time required to prepare them for use. But for racers planning to employ proprietary "trick" configurations, rough-machinings may leave too few options.

Bare heads are essentially ready-to-run once you drop in valves, springs, rocker studs, rockers, and a few other parts. The only machining needed is to grind the valves and seats for a proper, efficient seal.

Last, you have the option of fully assembled heads, which are the easiest, logical choice for many enthusiasts. Assembled heads, because they're fully machined and include valves, springs, and other parts, are ready to bolt on right out of the box, but they do cost more. Despite their premium price, however, they still tend to be less expensive than any of the other choices (of course, the parts used in the heads may not be what you would have selected on your own).

These ports have been ported and it's possible to see most of the valve quite easily, which indicates they flow very well. Note the D-shape of the late-model production aluminum heads' exhaust ports.

not extend to the bottom of the guide—it's interrupted—to prevent excessive oil consumption.

Simple iron guides, however, are actually a great solution for any application, except professional-level competition engines. Aside from being economical, they will provide excellent wear.

Guides should be replaced (or relined) whenever cylinder heads are rebuilt. Some machine shops, however, are fond of knurling the valve guides, rather than replacing or relining them, since the process is less expensive. But knurling actually digs up the inside of the guide and will often result in accelerated guide wear when the engine is put into operation.

Combustion Chambers

A cylinder head's combustion chamber is where the combustion process takes place, and its size and shape have a great deal to do with how efficiently the burning process takes place.

In the old days, combustion chambers were large—as much as 76 ccs during the 1970s. As GM learned the importance of an uninterrupted combustion process, it learned that domed pistons were inefficient. In order to switch to flat-topped or dished pistons (which don't obstruct the flame-front travel through the cylinder), the size of the combustion chamber had to be reduced to maintain a reasonable compression ratio for good power. As a result, many production Gen-I and Gen-II heads today have chambers as small as 72 ccs, while many race heads have unfinished chambers that are even smaller!

This may pose a problem for someone trying to build a high-performance street engine because the small-chambers of late-style performance heads may result in excessive compression, which would likely result in detonation when running pump gas.

In addition, the shape of the chamber has changed considerably from the half-moon-style chambers of old to the predominantly kidney bean–shaped chambers of today's heads. This was largely due to the move to flat-top or dished pistons, which do not inhibit flame-front travel the way domed pistons do. Lacking domes to keep compression up, the chamber size had to be reduced. According to various motorsports engineers, as the chambers shrank in size, they were reshaped to ensure fuel remained atomized (mainly for carbureted competition engines) and to help direct exhaust gases toward the exhaust

Valve Angles—New Slant on Performance

Production small-block cylinder heads have used a 23-degree valve angle since the engine first debuted in 1955. The angle is measured from a line perpendicular to the block's deck surface.

GM's Motorsports Technology Group discovered in the early 1990s that valves at shallower (lower numerical) angles produced better flow than the stock 23-degree valves. One of the first heads to feature a shallower valve angle was the 18-degree head for use on NASCAR Winston Cup engines. GM also created a 15-degree racing head and later employed 15-degree valve angles in the Gen-III production heads.

One aftermarket cylinder head manufacturer, Trick Flow Specialties (TFS), uses other-than-stock intake valve angles in its Chevy small-block heads. These are known as the Twisted Wedge heads. The TFS Twisted Wedge heads have the intake valve stood up 10-degree straighter at 13 degrees. It should be pointed out, however, that unless modifications are made to the valvetrain, geometry differences can lead to premature guide wear.

In addition to the angle of a valve tilting toward the intake manifold or away from it, valves can also be canted toward the front or rear of the motor. Typically, such designs are only used on racing engines, where, again, increased guide wear is not of concern, since the engine will likely be torn down and rebuilt—including valve guide replacement—long before any detrimental effects occur.

valve during the exhaust stroke for better cylinder evacuation.

Another head characteristic related to chamber design is the angle of the spark plugs. Small-block heads are often referred to as "straight plug" or "angle plug." Aside from the obvious—that straight plug heads have spark plugs that stick straight out of the head, while angle plugs stick out at an angle—the difference is actually designed to alter the position of the spark plug tip in the combustion chamber for better combustion efficiency. Many aftermarket performance heads are actually available in either form. Though both designs work well, evidence suggests that angled plugs (which point the tip toward the exhaust valve) provide for a marginally better combustion process; however, angled spark plugs may interfere with exhaust headers or other accessories.

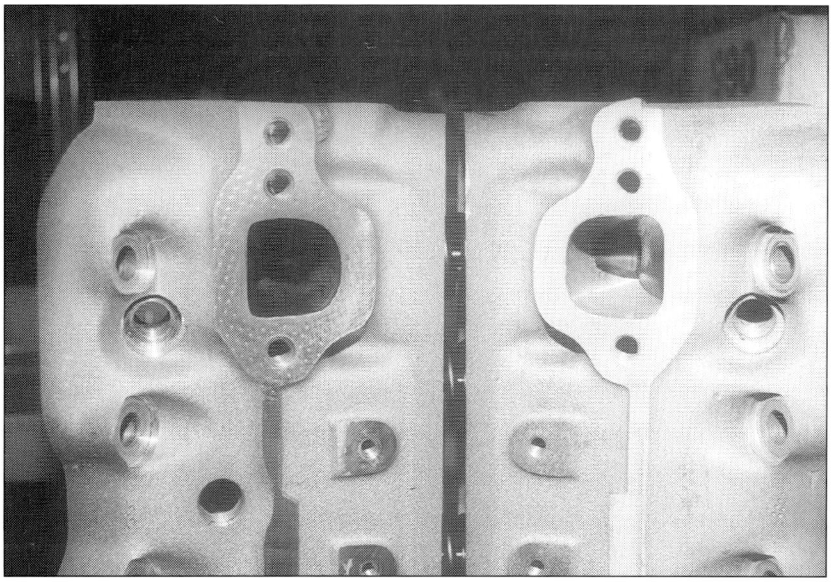

A production head is on the left, while a Callaway CNC-ported head is on the right. Note the considerable increase in size of the Callaway head.

Cylinder Head Materials

All first-generation TPI engines featured cast-iron cylinder heads, except for the L98-equipped cars that sported aluminum cylinder heads. LT1 and LT4 Gen-II engines feature aluminum heads, as well, except for the Impala SS LT1, which has iron cylinder heads. The Gen-III LS1 and Vortec 4800 and 5300 engines also use aluminum heads, while the heavy-duty Vortec 6000 features an iron head. Aftermarket heads are available in both iron and aluminum; some heads are even available in either material.

The particular material from which a given cylinder head is made doesn't affect airflow through the head, but it does affect power production. Aluminum heads run cooler, which allows the ignition process to begin later when the piston is closer to top dead center on the compression stroke, because detonation does not set in as early as with hotter-running iron heads. This allows the air/fuel mixture to be more compressed before it

Aside from the rough finish, this photo shows odd, unnecessary bumps in the port that only serve to slow the flow of exhaust gases.

Aside from the flow aspect of cylinder heads, you also have to think about the valve guides, seen protruding here from the valve spring seat area. Several types of guides are available.

is ignited; it burns more easily, more thoroughly, and hotter so more power is produced. Aluminum heads also shave considerable weight from an engine.

Because of their softer metal, aluminum heads do require valve seat inserts, some form of valve spring seat cup, and also screw-in rocker studs.

Whereas iron heads are cast, and therefore have cylinder-to-cylinder variations in volume and shape of the combustion chambers and ports, most aftermarket aluminum heads are machined by Computer Numerically Controlled (CNC) equipment. CNC machining ensures that each cylinder's combustion chamber and intake and exhaust ports are exactly like those of every other cylinder's chamber and ports. This equality helps equalize power production from cylinder to cylinder, which results in a balanced, smoother-running engine.

Cylinder Head Recommendations

Unless there is some overwhelming reason, such as competition class rules, that require a cast-iron cylinder head,

Like the ports, the valves have an immense effect on flow. The center two valves are stainless steel, high-performance valves, while the ones on the ends are stock.

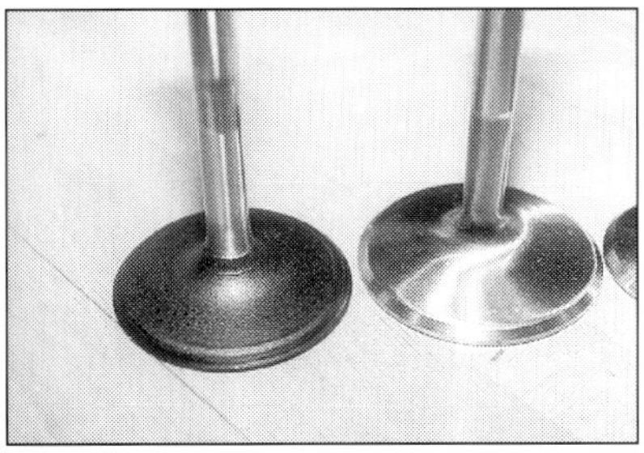

Note the difference in finishes from the stock valve (left) to the stainless valve. The smoother surface actually has a noticeable effect on flow, as does the multi-angle valve job.

Of particular interest in this shot is the thickness of the neck of the stock exhaust valve (center, left) versus the high-performance "necked-down" stainless valve (center, right). The impact on flow should be obvious.

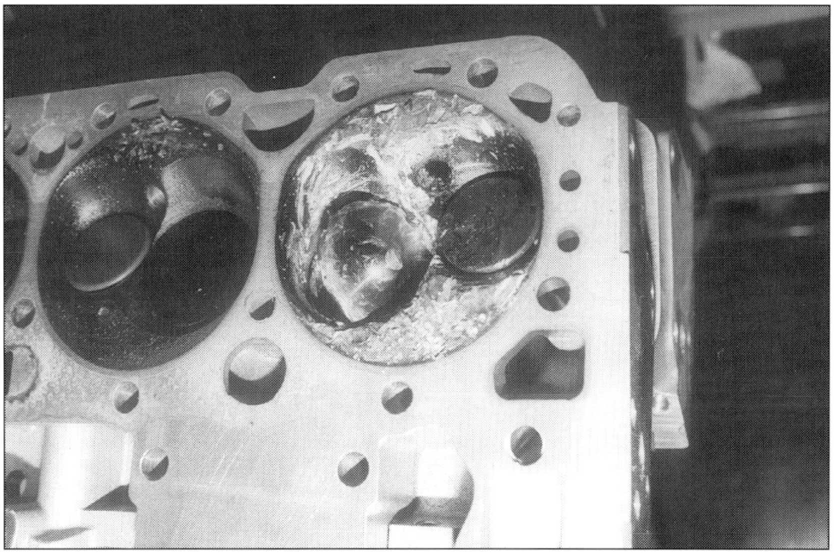

Ouch! This is what happens when you "drop a valve." Not a pretty picture, is it? The valve hit the piston and broke, the piston hammered it around a while in the chamber digging all the little divets, then pounded the valve head into the bowl.

Collateral damage from the previous dropped-valve shot. Many of those divets from the first cylinder got knocked back into the intake manifold and made their way to other cylinders, causing the damage you see here in each chamber. The cylinder walls were also severely damaged.

aluminum heads are generally preferable, and given the modest difference in price, there's even economic sense to running aluminum heads.

If your current cylinder heads have a lot of miles on them, or require extensive machining to configure them the way you want them, you may find that replacement heads may be a better buy than paying for lots of machine work. It's also often quicker to order a replacement set of heads than to deal with pulling your current heads, getting the machine work done, and getting them back. So, if you're in a rush, replacement heads may make even more sense.

For high-performance street use, intake port sizes of 180 to 199 ccs tend to work well, depending upon other components, like the camshaft and induction and exhaust systems.

Autocrossers will often find that smaller ports—under 180 ccs—tend to improve low- and midrange response, and thus provide power where it's needed when darting in and out of cones.

Again depending upon other engine components, road racers and drag racers will often find that heads in the 190-plus range work best.

CHAPTER FOUR

TPI INDUCTION SYSTEM

When we talk about tuned port injection engines, we're really talking about engines with the TPI induction system—an electronic fuel-injection system that highly efficiently mixes fuel with air to make power.

It's a TPI engine's induction system that makes it different from non-EFI and EFI Chevy engines alike. When the TPI induction system debuted it was a definite advance. Until that point, fuel injection had always been implemented as more or less an electronic carburetor: The injectors mixed fuel with the incoming air before entering the intake manifold. This resulted in a relatively heavy air/fuel mixture flowing through the intake. Because intake manifold runners are generally of unequal lengths, the air/fuel mixture flows more easily to some cylinders than to others, which results in an uneven distribution to the cylinders, which means power isn't produced evenly from cylinder to cylinder.

But TPI changed all that. Instead of flowing a heavy air/fuel mixture through the intake, only air flows through it, which is much lighter and flows more easily through varying shapes and sizes of passages. Fuel is added by an individual fuel injector for each cylinder, just ahead of the cylinder head intake port, aimed at the intake valve for a clear path to the cylinder.

In addition, TPI resolved the unequal runner problem, so each cylinder had the same length of induction tract as every other cylinder, which further helped balance power by balancing the intake airflow through the induction system to the intake valve of each cylinder.

Because the induction system controls the volume and quality of both the air and fuel reaching your engine's cylinders, it is largely responsible for the overall performance of your engine. By altering the volume and quality of air and fuel delivered to the cylinders, you

Your engine's induction system is responsible for getting air and fuel into the engine to burn to make power. As you probably expect, there are lots of ways to upgrade stock TPI induction systems, as this engine shows with its ACCEL SuperRam plenum and Vortech supercharger. *Vortech*

Callaway Cars' Twin Turbo L98 Corvettes were a factory option (though Callaway assembled the cars). The stock TPI units were ported and upgraded, and the twin turbos and intercoolers were added, of course.

alter the engine's ability to produce power at different engine speeds. Warm air or air that doesn't flow smoothly to the cylinders reduce power: Warm air is less dense than cool air, so there is less oxygen with which fuel molecules can bind, while poor flow quality results in air stumbling in the intake instead of being highly focused and aimed toward a specific cylinder. So the cylinder may not receive the same volume of air and fuel as would a similar cylinder in an engine with a more efficient induction system.

Air Induction

Understanding Airflow

Before you embark on a project to upgrade your engine's induction system, it helps to remember a few basic rules and principles about airflow.

The number one principle to keep in mind is that an engine is really nothing more than an air pump: It draws air in and pumps air out. The more efficiently it can do both of these (at a given crankshaft speed), the more power it can produce. This chapter deals with the first half of the equation—getting air into the engine—but you really can't make upgrades to your induction system without examining the flow abilities of your car's exhaust system too.

It also helps to keep in mind that we aren't necessarily talking about maximizing the volume of air that can flow in and out of the engine, but also the air temperature and flow characteristics of the air flowing in and out of the engine under the driving conditions that your engine will face. What this means is that bigger isn't necessarily better. If it were, we would all run sewer-sized exhaust pipes and equally large intake ports and runners, which would obviously flow more air than smaller pipes and ports. The problem with gargantuan-sized passages

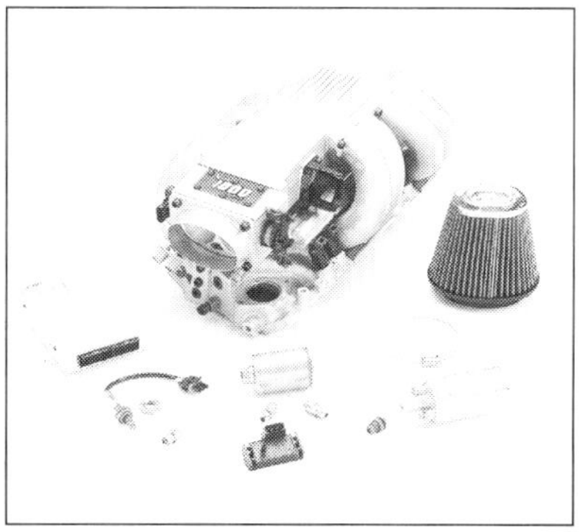

Complete aftermarket TPI systems are available for retrofit onto non-EFI cars, or as upgrades for stock TPI cars. This is ACCEL's StreetRam, which features oversized runners, a ported plenum and base, and a 1,000-cubic feet per minute throttle-body assembly, plus a fuel pump, ECM, air filter, and everything else needed for a clean, easy install. *ACCEL*

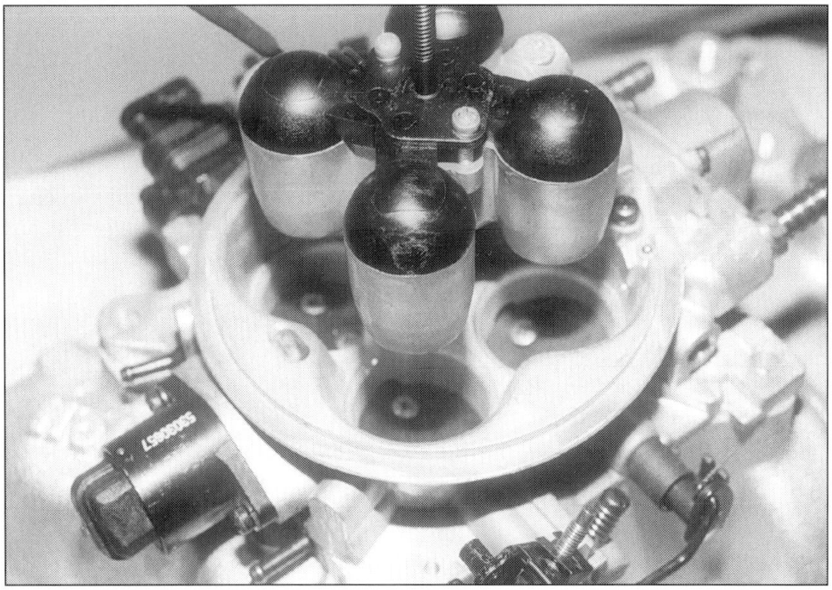

GM's initial forays into the world of EFI were in 1982 with throttle-body injection units extremely similar to this Holley Pro-Jection TBI unit. The large cylinders above the venturis are the fuel injectors. Electronics control how much fuel gets squirted.

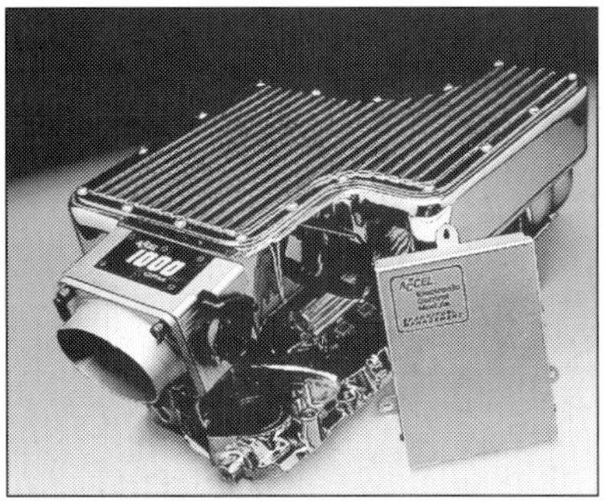

Another upgrade for the TPI system is ACCEL's SuperRam system, which is also available as a complete kit for installation on non-TPI cars or as an upgrade for TPI cars. The SuperRam's runners are shorter than the stock TPI units, so they flow better at higher rpm, without giving up much down low. *ACCEL*

is that, while they have the capacity to flow vast quantities of air, your engine doesn't always require vast quantities of air, especially at lower rpm. And at low rpm, where the engine only draws a little air in, the air moves slowly through those massive ports because it doesn't stay focused.

Here's a quick experiment to help you understand the effects of port sizes on airflow. You'll need a regular drinking straw and the cardboard tube from a roll of paper towels. First suck air in through the straw. Notice how the air rushes into your mouth quickly. Now try sucking air through the cardboard paper towel tube. Your lungs are still only going to draw in the same volume of air, but the flow velocity is greater through the straw than the paper towel tube, and faster airflow velocities can create a ramming effect that ushers more air into the cylinders, especially at lower engine speeds.

So, smaller ports make for higher airflow velocities, which helps pack more air and fuel into the cylinders. Combustion efficiency is increased, so the engine produces more power—to a point. As engine speeds increase, those small ports pose a restriction, limiting how much air can flow into the engine during the intake stroke. Imagine trying to run breathing only through that tiny straw—you wouldn't be able to inhale the oxygen-rich air your body would need to sustain the activity. That cardboard paper towel tube wouldn't pose much if any restriction, however, so you would be just fine breathing through it.

As with the diameter of your induction system passages, the lengths alter efficiency too. More breathing exercises make this easier to understand. The straw you breathed through a moment ago was relatively long—several inches—and while that the air was at a high velocity, you couldn't breathe quickly through it. Now simply pucker your lips, leaving an opening roughly the size of the straw and breathe in. While the opening through which air flows into your mouth is (roughly) the same size, it's easier to breath because the length of the port is shorter—just the thickness of your lips, instead of the several inches of straw. Though it's still not ideal, it is conceivable

GM revised the TBI system later in the 1980s for use on its truck line and other passenger cars. TBI is less accurate and less efficient than TPI-type multipoint injection systems. TBI is far more efficient than a carburetor and far less costly than TPI, however. GM continued to use TBI through 1995, attesting to its reliability and efficiency.

In 1990, Chevrolet introduced the world to a TPI system on steroids: the mighty ZR-1's twin-runner TPI LT5 engine. The engine operated on one runner per cylinder under normal driving, but when engine demand warranted, butterflies opened the second venturis for increased airflow and power.

that you could breathe with only minimal difficulty through your lips in the described manner.

So the volume of air you (or your engine) can breathe is directly related to how long or short the runners are: Shorter runners flow more air in a given period of time, while longer ones become a restriction at high engine speeds.

The obvious comparisons to these two examples are the long-runner first-generation TPI induction systems and the short-runner LT1/LT4 systems. The port diameters are roughly the same size between the two intakes, but the TPI runners are some 18 inches long, while the LT1s are about 3 inches long. In terms of how this effects engine performance, the TPI engine begins having noticeable breathing trouble at about 4,500 rpm, while the LT1 can continue until nearly 6,000 rpm before power drops off appreciably. But because each relies on modest port diameters, each

The 1992 LT1 had a redesigned, short-runner TPI system that flowed much better at higher rpm and, thanks to other changes in the new Gen-II engine, didn't give up much power during low-rpm driving.

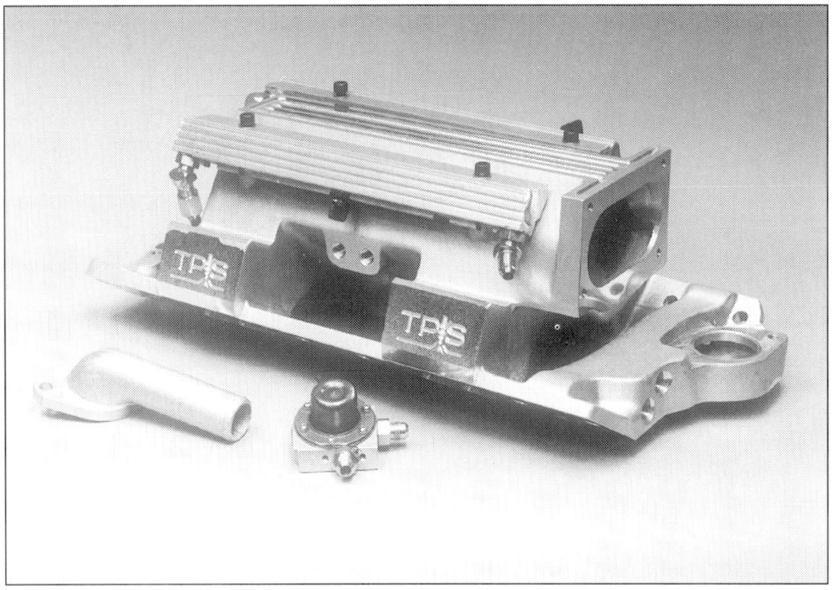

TPI Specialties created an LT1-like intake for first-generation TPI engines, called the Mini-Ram. With its extremely short intake runners, high-rpm breathing is greatly improved. Because torque does fall off a tad, it works especially well with large-displacement "stroker" motors, or with engines bolted to a manual transmission. *TPI Specialties*

produces good power at moderate engine speeds.

Of course, this discussion has been greatly simplified. In actuality, air temperature—and thus the *density* of the air—is as important as, if not more than, flow velocity, because density of the air is a measurement of how many oxygen molecules are contained in a given volume of air. Because fuel molecules need oxygen molecules to bond to, more oxygen molecules in a given volume of air means more fuel molecules can be supported by that specific volume.

Compromises

There are a few interesting notes worth knowing about the factory TPI and LT1 intakes. First, the long-runner TPI intake was originally designed to complement the 305-cubic inch small-block at a time when the future of GM's 350-inch small-block looked very bleak, thus the intake's runner lengths and port sizes are tuned to

Chevy's latest step is the Gen-III engine line, which includes the all-aluminum LS1 used in the Corvette, Camaro, and Firebird, plus the Vortec 4800, 5300, and 6000 cast-iron truck engines. This particular LS1 has been heavily modified by Callaway Cars for use in its C12 exotic sports car and delivers roughly 450 horsepower, yet it still retains the stock Gen-III TPI composite intake and monoblade throttle-body assembly.

Cold Air Induction

Early on in this chapter we discussed the notion that an induction system isn't just supposed to provide a given quantity of air, but rather a supply of high quality air. One way to improve the quality of air reaching your engine is to make sure that air is as cool and dry as it can be, because cool air is more dense than warm air, which allows it to hold more fuel. Dry air, likewise, isn't encumbered with carrying water vapor, which allows it to carry more fuel. The more fuel the air can carry, the hotter the combustion process will be, so the more power your engine will make.

The best way to supply cool air to your engine is to incorporate some form of air scoop to funnel air to the induction system. A hood scoop is the obvious choice, but aftermarket systems that mount a scoop below a vehicle's front fascia, behind the grille or even behind an empty headlight bucket, may be easier and cheaper to incorporate—and sometimes even more effective.

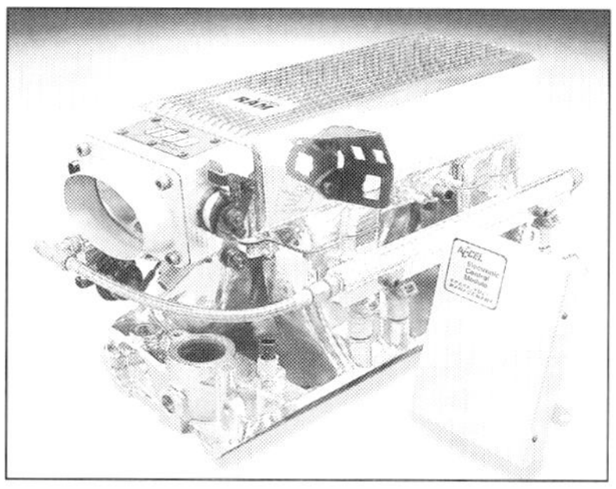

One of the nice features of the TPI multi-point fuel-injection design is that it can be modified and adapted to other engines easily. This ACCEL SuperRam system for big-block Chevys is a great way to trick-out a custom big-block installation or upgrade a carbureted or TBI-equipped engine. *ACCEL*

maximize efficiency with the small-inch engine. This explains why a dual-catalyst-equipped TPI 305 made 230 horsepower (0.75 horsepower per cubic inch), while a similar 350 made only 15 horses more (0.7 horsepower per cubic inch). It also explains why L98/B2L 350-inch TPI engines respond so well to oversize aftermarket intake runners and ported manifolds—the engine was being choked off by the stock intake system.

A second note concerns the short-runner LT1/LT4/L99 intake. It, like the TPI runners, was a design compromise; in this case it was sized to fit under the fourth-generation Camaro/Firebird radically sloped and forward-reaching windshield. As such, the runners are likely a bit shorter than Chevy engineers may have preferred because the intake does not perform well at low engine speeds. Anyone who has driven a car equipped with an LT1 is likely aware of the surging effects below about 1,500 rpm. Their

Here's a unique, custom twin-plenum cross-ram TPI-style induction system fabricated for use on this Camaro's 572-cubic inch, all-aluminum, 700-horsepower big-block. Chevrolet's Raceshop built the car as a "what if" project to show what muscle cars might be like if the company still built big-blocks for passenger cars.

Lingenfelter Performance Engineering crafted this sheet metal intake and used an LT5's throttle-body assembly to feed a 427-cubic inch Chevy SB2 small-block. The engine made more than 600 horsepower and incredible torque!

Getting back to reality, one of the easiest improvements you can make to your TPI engine is to replace or upgrade its air filter. This is a stock-like pleated paper filter, which is very restrictive.

cause: The runners are so short that pulses from one cylinder disrupt the airflow into another, so the affected cylinder makes substantially different power from one firing to the next. Longer runners would isolate each cylinder from the others, minimizing any flow disruption.

Back to the Beginning

So far, all we've covered are the parts that come after the throttle body assembly, but that only tells half the intake story. Airflow-wise, the induction system is divided into two separate but interdependent air intake systems: everything from the throttle body forward and everything after it.

The first half of the induction system has one simple function: to supply the engine with a sufficient volume of clean, filtered—and preferably cool—air to satisfy the engine's demands. In general, stock systems present a fairly considerable restriction to airflow, which limits how much power the engine can make, regardless of how efficient the second half of the induction system may be. In short, if you can't get enough air to the intake runners, it doesn't matter how much air they can flow because they won't be getting enough to fill the cylinders. If you increase the volume of air that can reach the engine, such as by increasing the size of the ducting to the throttle body, you reduce the restriction, so power will increase. Note that power increases will only be realized to the point where the engine no longer encounters a restriction; once the engine's needs are fully satisfied with a given ducting design, a larger design won't meet them better to produce more power. In fact, as with intake runners, larger intake ducting makes for slower-moving air, which can actually cost you power. So it pays to make careful, conservative decisions when upgrading your air intake ducting and throttle body assembly.

The exception to all this are forced induction systems, such as superchargers, turbochargers, or even just ram air induction systems, which don't wait for an engine to "request" air but rather force the air down the engine's throat.

Considering Airflow Upgrades

For any changes to be successful, you need to plan them carefully to match your driving conditions, which involves considering everything from vehicle weight, transmission type, rear axle gearing, expected engine operating speeds, to your driving style, and—of course—engine size, among other factors.

Obviously, smaller displacement engines, like 305s, require less air than larger ones, so the changes you make for small engines will differ from those you make for larger ones. What would be considered a mild upgrade for a 350- or 383-inch engine may be a radical change on a 305—too radical. Your changes have to be sized to suit the engine you're dealing with, and when in doubt, be conservative, think small.

Your transmission has a very direct impact on the power demands of your engine. Most manual transmissions have closely spaced gears, which result in less of an rpm drop from one gear to the next. That being the case, power can be made in a narrower—and generally higher—rpm band. Automatics, however, typically result in a fairly large rpm drop when changing from one gear to the next, so power curves need to be broader to prevent the engine from faltering after a gear change. Put in specific terms, shorter, fatter intake runners work better with manual transmissions, while longer runners tend to favor automatics.

Vehicle weight is a factor, because the heavier a car is, the more torque you need to move it and the more you need that torque spread out over a broader rpm band to accommodate various driving situations. This translates into a need for longer runners than a lightweight car requires.

Rear axle gearing and expected engine operating speeds go hand-in-hand: low (high numerically) gears raise engine speeds. If you expect to be running your engine to 6,000 rpm, long-runner intakes are going to choke the engine off. On the other hand, if you're trying to build a motor with tons of low-end grunt— (say for a tow vehicle)—you'll want those long, small-diameter runners to keep air velocities up, which improves torque.

An improvement is switching to an oiled-cotton filter, like this one from HyperTech. Not only does air pass through the cotton element more easily, but the oily cotton fibers trap dirt and other contaminants better. The filter is even cleanable and reusable, so it could be the last filter you ever buy.

Even better is this oiled-foam filter from ACCEL. Air flows through the foam element virtually unrestricted, yet the oiled foam still does an outstanding job trapping debris. It, too, is cleanable and reusable.

Faster Filters

None of this talk of runner sizes and shapes does any good, of course, unless you can get enough air to the intake runners through the air filter, ducting, and throttle body. In general, however, unless you hand fabricate your own intake ducting, you will be at the mercy of General Motors and the aftermarket for the parts that are available to you. One of the biggest restrictions in your induction tract is your air filter. Stock folded paper-

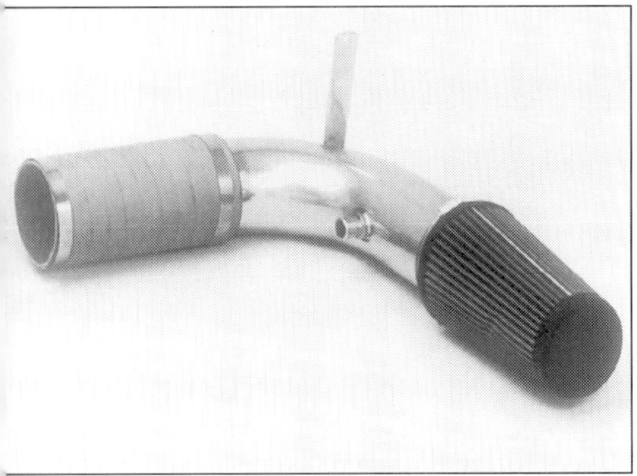

Once the air gets through the filter, it needs to efficiently flow to the engine, and that requires efficient intake ducts, like this stainless steel, mandrel-bent pipe for fourth-generation Camaros.

element filters present a significant restriction to airflow. Two types of high-flow replacement filters are available, each of which provide noticeable improvements: oiled gauze-type filter elements and oiled foam-type elements.

The gauze-type elements feature smaller, more numerous filter fibers, which both increase filtration characteristics and decrease flow restrictions through the filter. As air flows through the filter, the smaller fibers don't disrupt the airflow as much. Special oils applied to the filter make the fibers tacky, so any particles in the air stick to the fibers like glue. A side benefit of these filters is that they can be cleaned and reused.

Foam-type elements aren't pleated like paper and gauge filters, but are generally thicker and, like the gauze filters, get coated with oil to attract dirt and dust particles. The real beauty of the foam filters are that they are extremely porous, which allows air to flow through virtually unrestricted, yet the filter's surface area increases its effectiveness at stripping contaminants from the air flowing through it. Again like the gauze-type filters, foam filters can be cleaned and reused, so you'll buy it once and it will last the lifetime of your vehicle—or longer.

Get Your Ducts in a Row

After the filter we need to consider tubing or hoses to route the clean air to the throttle body assembly. Factory systems are often inefficient, because, in addition to flowing air, they are designed to reduce engine noise. To control sound, the ducting often contains resonance chambers that are tuned to reduce specific frequencies or ranges of frequencies. Unfortunately, these chambers also disrupt airflow, because they send some air into a dead-end chamber. Then it bounces back into the ducting, colliding with in-rushing air. Removing these resonance chambers is a quick way to increase airflow through the ducting.

Another problem many production intake ducting systems face are design limitations imposed by the need to be mass-produced, or to fit into a given engine compartment, for instance. Such restrictions often dictate tight bends or collapsed sections to provide clearance

You don't have to be an engineer to see that this Impala SS induction system ducting isn't exactly a straight-shot to the throttle body assembly. The large, rectangular box is a sound-absorbing chamber used to tune out undesirable intake noise, but it also greatly disrupts airflow. The triangular piece is simply a dress-up item.

LS1-equipped Camaros also have an intake tuning chamber that upsets airflow. It's the large H-shaped apparatus between the air filter housing and the mass airflow sensor.

for other equipment, but these tight bends and reduced sectional areas kill flow through the ducts. Replacing the ducts with less-compromised designs will increase flow.

Finally, there's the throttle body. Not to be confused with a throttle-body injection EFI system, a TPI throttle body only controls the airflow into the engine by means of a throttle blade or blades, just like those of a carburetor's base plate. A throttle-body injection system not only controls airflow into the system, but also mixes fuel into that air—similar to a carburetor, but more precise. Stock throttle body assemblies on TPI and LT1/LT4 engines use twin throttle blades, while the LS1 and Vortec engines use a single, large, round throttle blade. Several aftermarket companies produce larger-diameter twin throttle body assemblies for the TPI and LTx engines,

In addition to tuning chambers, stock intake ducting is often compromised to make room for other equipment that needs to go underhood. Note the flattened duct along the radiator in this 1993 Camaro Z28's engine compartment. It's a severe impediment to airflow.

Forced Induction

Most of this chapter explores modifications and upgrades that are aimed at improving airflow into a naturally aspirated engine—doing what we can to help an engine suck in more air and fuel. But one of the quickest ways to dramatically increase airflow into your engine is to force-feed it air. Superchargers, turbochargers, and ram air systems blow air into the engine under pressure, so that when the intake valve of a given cylinder opens, more air and fuel enter the cylinder than it would naturally ingest, so the combustion process produces more heat, which means more power. In a nutshell, forced induction systems increase volumetric efficiency.

You'll find much more information about forced induction systems in chapter 9.

Aftermarket companies don't face quite the same design criteria GM imposes on its engineers, so the solutions can look pretty radical, like Callaway Engineering's "Honker" system for fourth-generation LT1 Camaros. It also makes a dramatic improvement in airflow.

and at least one company produces an oval mono-blade throttle body assembly that has the potential to flow significantly more air than a stock assembly.

Recommended Airflow Upgrades

High-Performance Street Driving

Street driving is uniquely demanding for an engine. An engine is asked to perform in a wide variety of conditions, from extremely slow-speed driving, including stop-and-go driving that may leave an engine idling for extended periods of time, to high-speed enthusiastic driving. These two situations have vastly different needs in terms of airflow.

In general, however, the bulk of all street driving—including high-performance street driving—occurs at lower engine speeds. And when engine speeds do extend into the higher ranges, they are only there briefly. Thus, it makes little sense to set up an engine to be efficient at mid- to high-rpm when it will spend 95 percent of its time at low- to mid-rpm.

With that in mind, longer intake runners will provide better torque, and runner diameters should be kept conservative to keep air velocities high. On a 305, your best bet may be to leave the stock TPI setup alone, except for port-matching to ensure smooth transitions from the plenum to the runners, from the runners to the intake base, and from the intake base to the heads. If the engine has been built up with a hotter cam, perhaps higher compression pistons, and better cylinder heads, it is likely that a set of oversize performance runners and a matching manifold base will increase airflow enough to meet the engine's needs without being too big. Siamesed high-performance runners are likely to hurt performance on a 305 (even a built-up one) more than they help, unless the vehicle is equipped with a manual transmission,

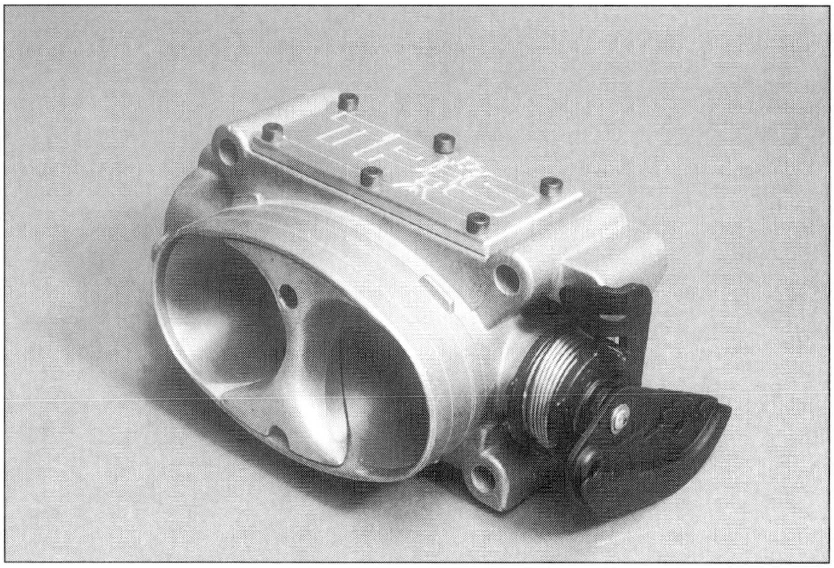

From the intake ducting, the air enters the throttle-body assembly. Don't confuse a multipoint injection system's throttle body with a Throttle-Body Injection (TBI) system. TBI systems meter air and mix fuel with it. Multipoint throttle bodies only meter air. This is TPI Specialties' ported 52-millimeter throttle-body assembly. *TPI Specialties*

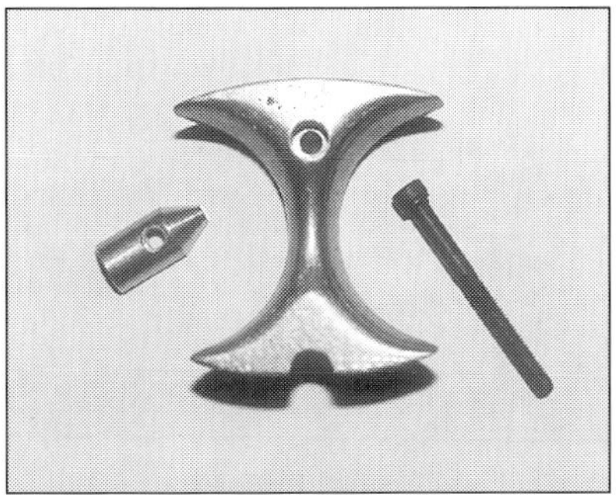

One of the best—and easiest—upgrades for stock throttle bodies is the addition of a flow directing air foil to clean up the blunt divider between the stock throttle body venturis. It takes about 10 minutes to install and increases flow by 36 cubic feet per minute, which can mean as much as nearly 12 horsepower.

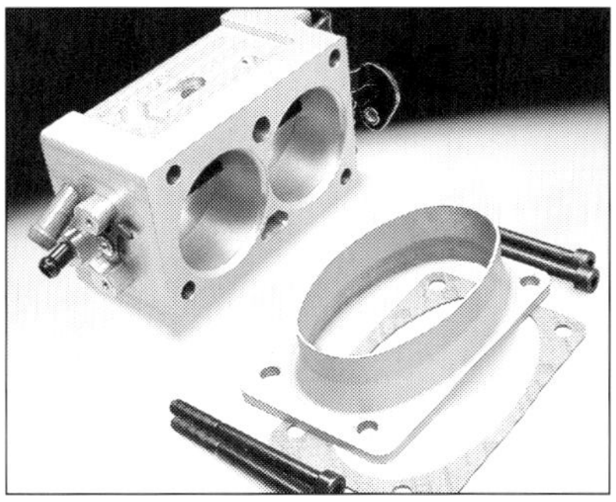

For stouter TPI engines, upgrading to a larger throttle-body assembly may be the only way to alleviate a restriction. This is ACCEL's 58-millimeter 1,000-cubic feet per minute billet throttle-body assembly, which will supply enough air to feed 600 hungry horses. *ACCEL*

The Vortec V-8s and Gen-III engines use similar monoblade throttle body assemblies. This is a Corvette's unit, as identified by the electronic throttle control.

which allows the engine to operate in a narrower, higher rpm band, where the siamesed runners may be of benefit.

Since we already know the TPI 350's production runners and base manifold were designed for the smaller 305, upgrading them to larger diameter runners and a matching manifold will benefit even a stock 350. Modestly built-up 350s, likewise, will work well with the oversize runners and intake, though, as with the 305, engines mated to manual transmissions may be able to make good use of siamesed runners, though low-speed torque will suffer somewhat, which is why they're not recommended for automatic-equipped vehicles, unless the engine features substantial performance upgrades.

Highly-modified 350s—including the LT1 and LT4—are generally better off with shorter runners. Essentially two options exist: ACCEL's John Lingenfelter

Because TPI systems monitor the throttle position (with a throttle position sensor, or TPS), it's critical that that sensor be properly adjusted. At closed throttle the voltage signal should be nearly 0, while at wide open throttle it should be nearly 5 volts.

Adding Fuel to the Fire

Depending on the modifications you make to your TPI engine, at some point it's likely that you will need to increase fuel delivery to the cylinders. There are actually two methods of achieving this goal: increase the flow rate or increase the line pressure.

Chevy gave us a simple example of the former method: It installs different fuel injectors on 5.7-liter engines than on its 5.0-liter engines; the 5.7-liter's injectors are capable of flowing more fuel—measured in pounds per hour—in order to feed its thirst for more fuel. The "larger" injectors actually have larger fuel outlets and are roughly comparable to running larger jets in a carburetor: The bigger holes let more fuel flow through.

The other method is to increase the pressure of the fuel within the fuel lines and injectors. By subjecting the fuel to a higher delivery pressure, more fuel flows from the injector during a given period of "on-time," versus a lower pressure and the same fuel system components.

While each method will provide more fuel for the combustion process, Myron Cottrell of TPI Specialties and Cottrell Racing Engines points out that larger injectors aren't needed until the fuel requirements of the engine and its operating parameters dictate a higher rate of fuel delivery than can practically be achieved through increased line pressures. In fact, Cottrell has used the 5.0-liter V-8's smallish injectors with much success in engines that deliver more than 400 horsepower, once fuel pressure was increased sufficiently. Cottrell also points out that the increased fuel pressure tends to create a better spray pattern from the injector, which leads to more even fuel distribution going into the cylinder, which eases ignition and promotes more complete, more efficient combustion for increased power. Lower fuel line pressures can result in fuel dribbling out of the injector nozzle in much the same way water dribbles from a garden hose at low pressure, and that leads to poor fuel distribution in the intake air charge, which can lead to localized rich and lean "pockets" in the combustion chamber. A rich pocket around the spark plug requires more energy to ignite and burns more slowly, whereas lean pockets don't burn well either, so the flame front may actually burn out, potentially leaving other pockets of air and fuel unburned within the chamber.

The easiest thing for most enthusiasts to do is to simply reuse whatever injectors they have, unless they're building a race engine that has more exact fuel requirements. An adjustable fuel pressure regulator and an accurate fuel pressure gauge can be used to increase pressure to the desired level (usually 42 to 45 psi).

One last note: Not all injectors are created equal. Some injectors have the ability to flow different volumes of fuel at the same line pressures. To maximize efficiency and balanced power production from cylinder to cylinder, you need injectors that flow the same volume at the same pressure. Also, injectors should have an even, diffused, conical spray pattern, which promotes even fuel distribution in the intake air charge.

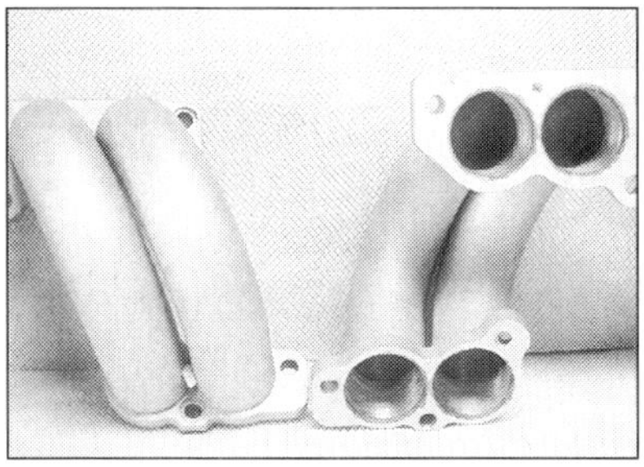

There are two basic types of replacement runners. These TPI Specialties' runners feature individual, oversized mandrel-bent tubes that are port-matched to the flanges.

Signature Series SuperRam intake system, or the stock LT1/LT4 intake. For first-generation TPI 350s, TPI Specialties manufacturer a virtual clone of the LT1/LT4 intake called the MiniRam, which is designed for use on first-generation engines that require a front coolant crossover. The SuperRam favors use with an automatic because of its

As in the old days, intake manifolds—including their runners—can make significant changes in an engine's ability to pump out power. This is Edelbrock's High-Flo aluminum manifold base and cast-aluminum runners, which feature oversize ports to compensate for the stock TPI ports, which were only designed to feed a 305-cubic inch engine.

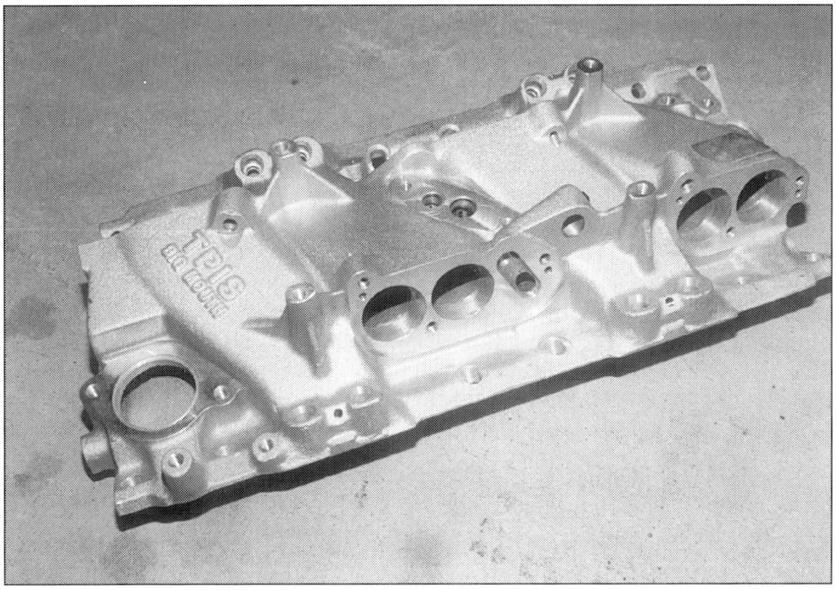

TPI Specialties' Big Mouth intake base has oversized ports for increased flow potential, and the ends of the ports are gasket-matched to minimize any flow disruptions as air transitions from the runners to the intake.

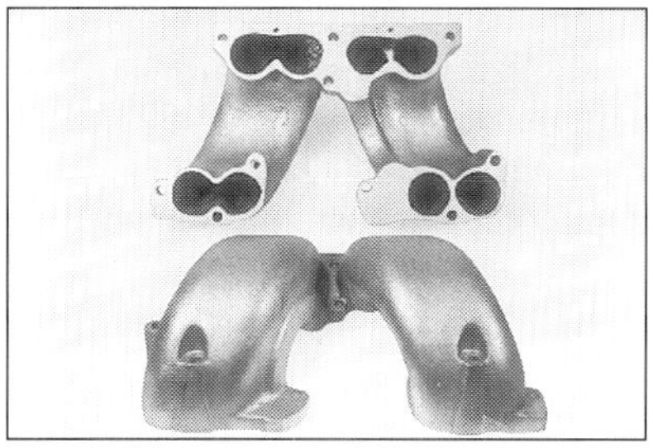

The second type of runner systems use siamesed runners, like these cast-aluminum units from SLP. The siamesed runners act like adding a bigger plenum and effectively shorten the runner length, thereby increasing breathing at higher rpm. *SLP*

medium-length runners that preserve more bottom-end torque, while the short-runner LTx/MiniRam intake will provide noticeably better performance with manual transmissions. Naturally, either intake can be used with either type of transmission with excellent results, depending on the specific engine and vehicle configuration.

A stroker small-block, like a 377- to 388-inch or larger motor will naturally make more torque than a 350, so it isn't necessary to worry about preserving torque with longer runners. Likewise, the air requirements of a large-displacement small-block are such that long runners will usually hurt performance at even moderate rpm. A mild big-inch small-block will typically respond well to a medium-length runner, such as ACCEL's SuperRam. Wilder engines will typically demand the airflow abilities of short runners, like those of the MiniRam or stock LT1/LT4 intake. In addition, stroker engines will also generally respond well to the increased flow capabilities of a performance throttle body assembly.

For every engine, stock to wild, minuscule L99 to massive 420-inch stroker motors, increasing the flow through the air filter and intake ducting will be of benefit. In addition, use of cold air induction systems will also help.

Drag Racing

Since drag racing calls for high-rpm power, you need to consider an induction system that will flow well under those conditions, and that means short- to medium-length runners. If you're on a budget and running a first-generation TPI engine, siamesed intake runners respond similarly to shorter runners, because they make the engine believe it simply has more plenum volume.

In general, unless your engine is a purpose-built maximum-effort race motor, you will likely be happier with the performance of a medium-length runner induction system, even if running a manual transmission and especially in heavier vehicles, because of the modest torque improvements it provides.

Autocrossing

True to the conception of autocrossing, its demands are much like those of street driving, though more concentrated toward the low- to mid-rpm range. Throttle response is a primary benefit in autocrossing, so longer, smaller-diameter runners

You can port a stock intake with a die-grinder and match the ports to the gaskets between the runners and intake for some flow improvements.

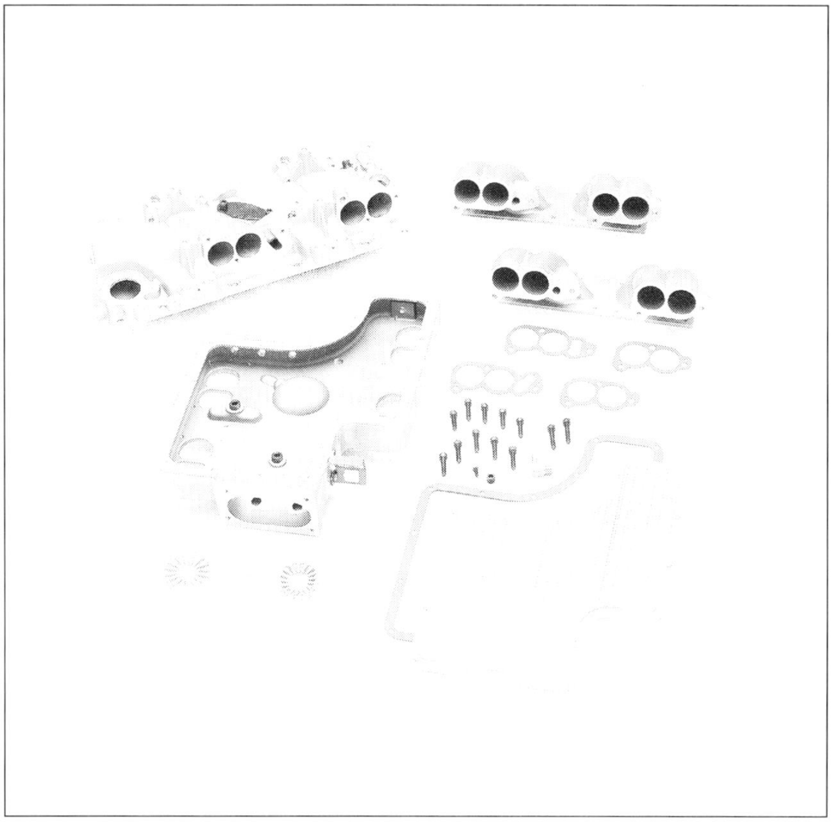

ACCEL's SuperRam intake system features a John Lingenfelter–designed plenum and medium-length runners that allow the intake to flow much better at higher rpm without killing low-rpm torque. The manifold base features oversize ports. *ACCEL*

are often the key. In fact, most first-generation engines will respond best in autocross competition with stock induction systems—especially 305-inch TPI engines. Gen-II engines (LT1, LT4, and L99) and any stroker motor will typically maximize usable torque if equipped with an intake with medium-length runners.

Road Racing

As with drag racing, road racing engines spend the majority of their time in the middle to upper end of the rpm spectrum. Unlike drag racing, however, road racing engines need to be docile enough to be lugged through tight hairpin corners at low rpm, without being choked off at high rpm on long straightaways, which would tend to indicate a medium- to short-length runner. However, given that most road race cars run manual transmissions, the shorter-length runner intakes—such as the TPI Specialties MiniRam or the stock LT1/LT4 intakes—offer the better compromise.

Air Induction System Components

The following section identifies and provides additional information about the various components of your engine's induction system.

Filters, Housings, Ducts, and Ram Air Systems

Stock air induction systems are often fairly restrictive, because GM's engineers have to meet certain requirements when designing a factory air induction system, including noise limits, fuel economy requirements, expected operating rpm, and even packaging sizes within a given vehicle. Fortunately, you aren't necessarily limited by these same design constraints when redesigning or replacing your engine's stock induction system. For instance, GM engineers equip Camaros, Impala SSs, and trucks with resonance chambers because many customers might find the unmuffled induction system offensively loud. But, if the sound doesn't bother you, you can eliminate the resonance chamber to clean up the airflow to the engine.

A stock air induction system usually starts with the air filter housing, which has an opening through which air can be drawn in. Ingested air then flows through a filter and into the ducts that ultimately funnel the air to the throttle body (either that of a throttle body injection unit or a TPI-style throttle body) and into the engine.

The one-piece Gen-III LS1 intake is made of a composite material, or what most of us know as plastic. The molded piece is very lightweight and has exceptionally smooth ports that would be almost impossible to create any other way than with plastic.

The underside of the intake clearly reveals the tall, narrow intake ports. Note the injector nozzle location at the ends of ports, which shoot almost directly at the back of the intake valves.

The front of the intake features a single, massive port for the monoblade throttle-body assembly. Note the unique O-ring-like gasket; every gasket (except the head gasket) on the Gen-III engines use similar rubber O-ring-style gaskets for exceptional sealing quality.

Unfortunately, there are often several problems with the stock air induction systems that cause them to choke an engine, to starve it for air. The first impediment is the air filter housing itself. Not only are the induction ports generally somewhat small (and therefore restrictive), but they typically draw air from a cramped, confined area. What air does make it into the housing must then pass through a highly restrictive paper-based filter, then find its way into the ducts that lead toward the throttle body. Once in the ducts, the ducts themselves often disrupt airflow with sharp bends, and some systems use the aforementioned resonance chambers that disturb airflow because some of the ingested air will flow into the chamber(s) then will swirl around and exit

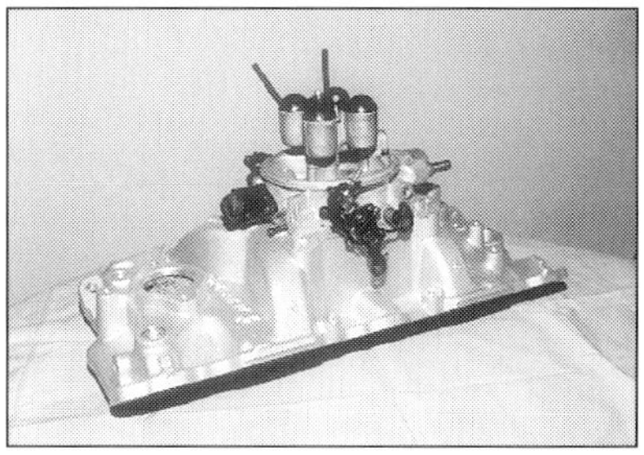

An alternative for someone looking to upgrade to EFI is to install a throttle-body injection (TBI) system, like Holley's Pro-Jection system, shown here atop Holley's Pro-Jection 4 aluminum dual-plane intake.

the chamber, reentering the duct and disrupting an otherwise smooth airflow.

Fortunately, there are numerous upgrades available that can completely eliminate any restrictions that the stock induction system presents. Easiest for the do-it-yourselfer is to replace the paper-element air filter with a high-performance oil-soaked cotton or foam filter. This simple change will immediately increase airflow to the engine, plus, because the aftermarket filters are typically serviceable (you can clean and re-oil them), you'll never have to buy another air filter again. Never run your engine without an air filter in place, no matter how tempted you might be by the thought of unrestricted airflow. It's simply not worth the risk. A single grain of sand can have devastating internal effects on your engine.

Several companies offer replacement ducts that eliminate not only the resonance chamber but other problems of the stock ducts too.

Another solution worth investigating for your engine is the installation of a cold air or ram air system. Several designs are available from a number of aftermarket companies, but each works essentially in the same way: At low speeds, the cold air induction systems allow the engine to more easily draw in air, while at faster speeds, air rushing into the cold air system scoop can actually be rammed through the filter, ducts, and into the engine, creating a slightly pressurized situation in the intake manifold that isn't unlike a supercharger or turbocharger's effect. In addition, because the air is generally drawn from outside the vehicle's engine compartment, the air is much cooler, and therefore is more dense

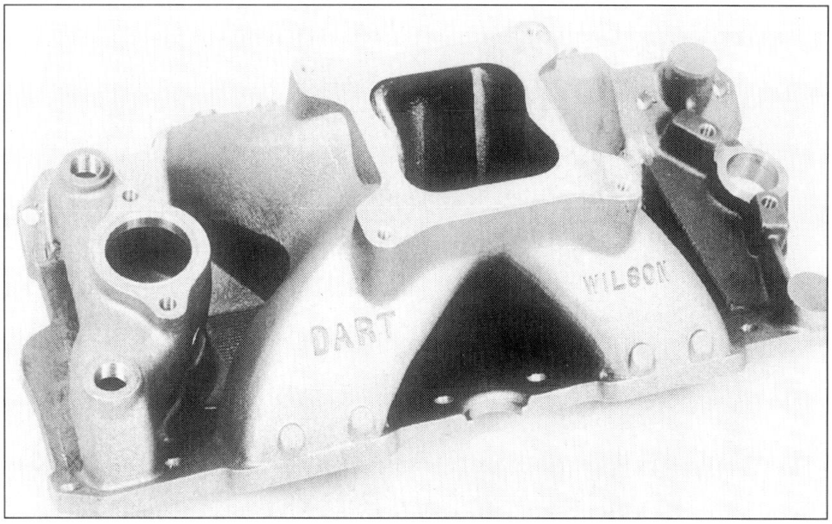

For racing purposes, it is also possible to use traditional single-plane style aftermarket intake manifolds, like this Dart/Wilson unit.

ACCEL's 1,200-cubic feet per minute billet aluminum throttle-body assembly can be fit to any standard four-barrel intake (Holley square-bore bolt pattern) to convert to a multipoint injection system. This is ACCEL's complete system with its own single-plane intake for big-block Chevy V-8s. *ACCEL*

For lower-powered applications, ACCEL offers an adapter to mount a TPI-style throttle -body assembly. A complete kit is available for conversion of non-EFI big-block engines. *ACCEL*

and will hold more fuel, which means a stronger burn cycle in the cylinder and more power.

Some cold air systems use a scoop mounted beneath the front of the vehicle, some use scoops and hoses mounted behind or in place of factory fog lights, and some even use unique cold air hoods with special hood scoops. Each system is effective, but all require some care on your part when driving in rain or other wet conditions because water may get in through the scoops and eventually get to the engine, where it could do harm.

Throttle Body Assembly

The throttle body assembly is simply a device equipped with a cable-activated or motor-driven butterfly valve to increase or decrease airflow into the engine. Carburetors have throttle bodies, so do throttle body injection units (big surprise there!), as do Tuned Port Injection systems.

The stock TPI throttle body can flow as much as 668 cubic feet per minute, according to TPI Specialties. Larger throttle bodies that can flow as much as 1,300 cubic feet per minute are available.

The only real modification that can be made to a stock TPI throttle body is to install an airfoil, which helps streamline the air entering the throttle body and can result in an increase of 41 extra cubic feet per minute over stock (for a total of 709 cubic feet per minute), according to TPI Specialties' flow-bench testing. Given that these tests measure only air flowing through the throttle body (as opposed to air and fuel, as you would have with a carburetor or TBI unit's throttle bodies), to compare these dry test numbers to the wet test numbers of carburetors, you need a carb about 10 percent larger to achieve the same flow rating since the fuel slows down flow through the throttle body. In other words, you would need an 800-cubic feet per minute carburetor to provide roughly the same flow as a 709-cubic feet per minute stock TPI throttle body outfitted with an airfoil.

With a 709-cubic feet per minute flow capability, the airfoil-equipped throttle body should be sufficient for engines producing 400 horsepower or so. More powerful engines will require larger-than-stock throttle bodies, such as 52-millimeter or 58-millimeter units, which flow 790 cubic feet per minute and 1,000 cubic feet per minute, according to TPI Specialties' tests. The 58-millimeter unit is acceptable for engines up to 600 horsepower. Above that, TPI Specialties offers a single-blade throttle body unit that flows 1,300 cubic feet per minute, and should be good for approximately 800 horsepower.

Throttle Body Airfoil

As mentioned in the previous section on throttle body assemblies, installing a simple airfoil to the inlet side of the stock TPI throttle body assembly can smooth the flow of air into the stock throttle body unit enough to increase maximum

Another good way to increase the efficiency of your induction system is to feed it cool air from outside the engine compartment, such as through a functional "ram air" hood, like this one from SLP. *SLP*

flow by 41 cubic feet per minute, for a total of 709 cubic feet per minute flow potential, and typically results in power gains between 10 and 13 horsepower.

The throttle body airfoil is a simple, effective solution that is strongly recommended for all B2L/L98, LB9, LT1, and LT4 engines. (LS1 engines cannot use—nor do they need—an airfoil because of their round, single-blade throttle body assembly.)

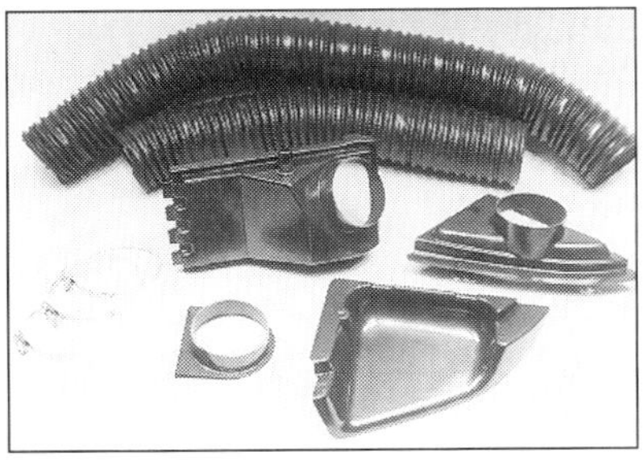

Cold-air induction kits can be fairly simple, like this one from Random Technology for fourth-generation Camaros, which consists of a scoop, a modified air filter housing, and some flexible hoses. *Random Technology*

Intake Manifold (Plenum, Runners, and Base)

Once air flows through the throttle body assembly, it enters the intake manifold assembly. Engines equipped with Throttle Body Injection systems use a traditional carburetor-style intake manifold, while 1985 to 1992 Tuned Port Injection engines feature a multipiece assembly that consists of an intake plenum with large curved runners that curl down to an intake manifold base. The 1992 to 1996 LT1, LT4, and L99 engines use a one-piece casting that integrates the plenum to the base with short runners. The pre-1999 Vortec engines use an air-metering throttle body atop a two-piece intake with a plastic plenum and aluminum base. The LS1 and 1999+ Vortec V-8s feature composite (that is, plastic) intakes. The LS1's intake features medium-length runners that curl up over the plenum to feed the cylinders on the opposite side of the engine, while the Vortec V-8s feature longer runners that aid torque production.

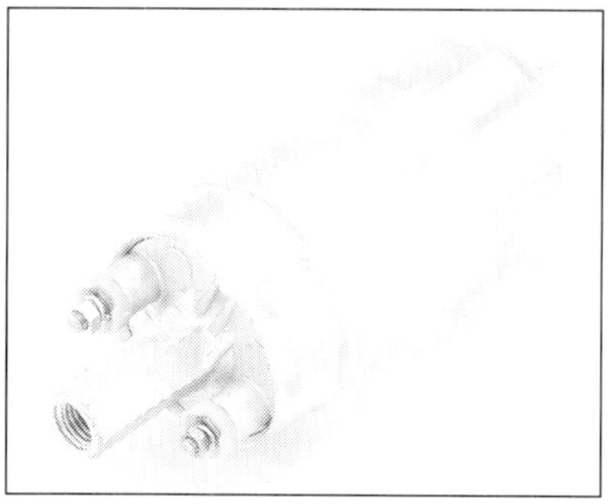

The other half of your induction system deals with fuel management, and for EFI systems a good, high-pressure fuel pump is critical. This ACCEL electric pump can flow 34 gallons per hour at 90 psi—enough to feed 870 horsepower! *ACCEL*

Because fuel injectors have such small orifices, a high-quality fuel filter is essential. Not only must the filter trap small particles, it must do so without creating a restriction to fuel flow. This is a stock, in-line AC canister filter on a 1998 Camaro.

Because EFI systems operate at such high fuel pressures, substandard fuel lines just won't cut it. These are the stock fuel hoses on an LS1-equipped 1998 Z28. The foil-like wrap not only insulates the lines from exhaust heat but also protects the lines from wear in case they come in contact with the ignition coils.

TBI systems can have their intake manifolds ported and port-matched to the cylinder heads to increase flow potential, or a variety of replacement intake manifolds are available for even greater flow—and power—potential.

TPI engines can also benefit from traditional porting and port-matching techniques, but they, too, have aftermarket parts available, including different intake plenums, larger-diameter or siamesed runner assemblies, and higher-performance intake manifold bases. Additionally, TPI Specialties offers an LT1/LT4-style MiniRam one-piece intake manifold assembly that greatly increases horsepower (though sacrifices some torque).

LT1, LT4, L99, and the TPI Specialties MiniRam intakes can also benefit from mild porting and port matching. The Chevrolet intakes feature a trapezoidal (nonrectangular) port shape that is narrower at the top of the port than at the bottom to improve flow characteristics. This general shape should not be defeated, otherwise flow will be less directed through the intake, and you may end up hurting performance.

Vortec V-6s and pre-1999 Vortec V-8s can have their aluminum manifold bases port-matched, plus the plenum area can be ported to match the composite plenum; however, the composite material cannot be effectively changed.

The LS1 and 1999+ Vortec V-8 composite intakes cannot be effectively modified, nor are replacement intakes available yet. And consider yourself warned: These composite intakes are extremely expensive, should you happen to ruin one attempting to modify it.

Cylinder Heads and Intake Valves

The final components of the induction system are the cylinder heads and intake valves; however, there are so many possible modifications and replacement parts available that discussion of these components is left to its own chapter, chapter 3.

Fuel System

The fuel system of your EFI engine is responsible for the storage and delivery of fuel to the engine. When burned, the fuel will produce power.

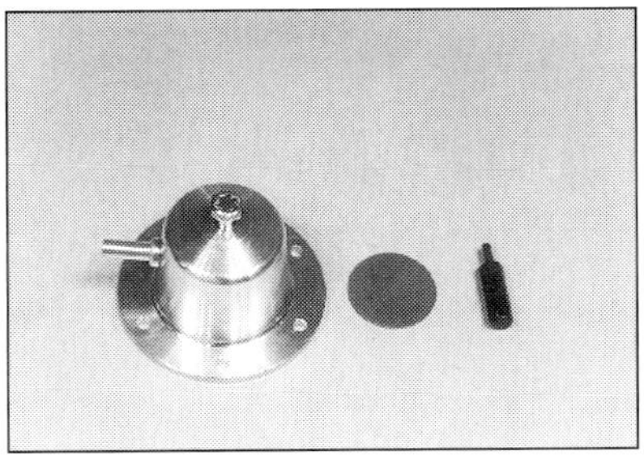

This is TPI Specialties' adjustable fuel pressure regulator, which lets you reduce or increase fuel pressure to suit your particular driving requirements, simply by turning a bolt. *TPI Specialties*

Chevrolet EFI engines have come in three basic flavors: Throttle-Body Injection, Central-Port Injection, and multipoint injection. Throttle-body injection squirts fuel into the air stream from above a carburetor-like device (stripped of any fuel-metering circuits) prior to entry into the intake manifold. Multipoint injection systems flow only air through the intake manifold, and fuel is injected directly into each individual port by a dedicated injector mounted in each runner. Central-Port Injection, which is used on Chevrolet's Vortec engine series, is similar to multipoint, except that the systems use only a single injector that sends fuel through flexible fuel lines to poppet valves mounted in each intake runner, just before the cylinder head.

In addition, multipoint fuel-injection systems come in two flavors: batch-fire and sequential-fire. Batch-fire systems, like the original TPI engines and the 1992–1993 LT1 engines, trigger all four injectors on a given engine bank at the same time. Thus, when the injector for cylinder 1 fires, so do those for cylinders 3, 5, and 7. Likewise, cylinders 2, 4, 6, and 8 all spray at the same time. The system is easier to engineer this way, but it's not as efficient. Sequential systems, on the other hand, monitor the crankshaft position more closely, to determine when a cylinder is nearing its intake stroke, then fires only the injector for that cylinder. The major benefit is improved slow-speed driving and reduced emissions output.

These differences aside, the fuel systems are actually fairly simple to comprehend and work with. A tank stores the fuel, an electric fuel pump delivers the fuel under high pressure through the fuel lines to a fuel pressure regulator, then to the fuel-injection system's injectors, be they on the throttle body, at each runner, or the central injector of the V-6 and pre-1999 V-8 Vortec engines. A fuel filter is mounted between the fuel tank and the engine to prevent debris from reaching the fuel injectors.

There are essentially two significant differences between the fuel system of EFI vehicles and those of old-world carbureted vehicles: the fuel pressure and the use of a fuel return line.

To ensure a fine, misting spray pattern from the fuel injector nozzle, EFI systems feature high fuel line pressures. Whereas a carbureted system will usually run between 5 and 8 pounds of pressure to the carb, a TPI fuel-injection system will deliver about 45 pounds of pressure to the injectors; the TPI fuel pump delivers more than that to ensure sufficient supply, but a pressure regulator reduces the pressure.

Because the fuel is delivered under such high pressure, a fuel return line is employed to route fuel back to the fuel tank. The return line provides the valuable benefit of minimizing fuel temperatures. Because anything subjected to high pressure increases in temperature, the fuel in the system could potentially become quite warm before being used, thereby reducing power. Second, because modern engine compartments are packed so tightly, they often are quite hot, which would naturally heat the fuel in any line that runs through the engine compartment. In both cases, however, the fuel is given ample time to cool off because the fuel runs back through the fuel line to the tank.

Several companies, including Mid America Designs, sell high-performance "balanced" fuel injectors that have been tested to verify a proper spray pattern. Note that injectors are available in different "pounds per hour" ratings to suit different engine displacement and power demands. *Mid America Designs*

Upgrades

Chevrolet's EFI fuel systems are plenty efficient to deliver fuel to high-performance engines. In fact, there are only a few recommended modifications.

To adjust fuel pressure you'll need a high-pressure fuel gauge like this TPI Specialties unit. It connects quickly and easily to a fitting on the fuel rail.

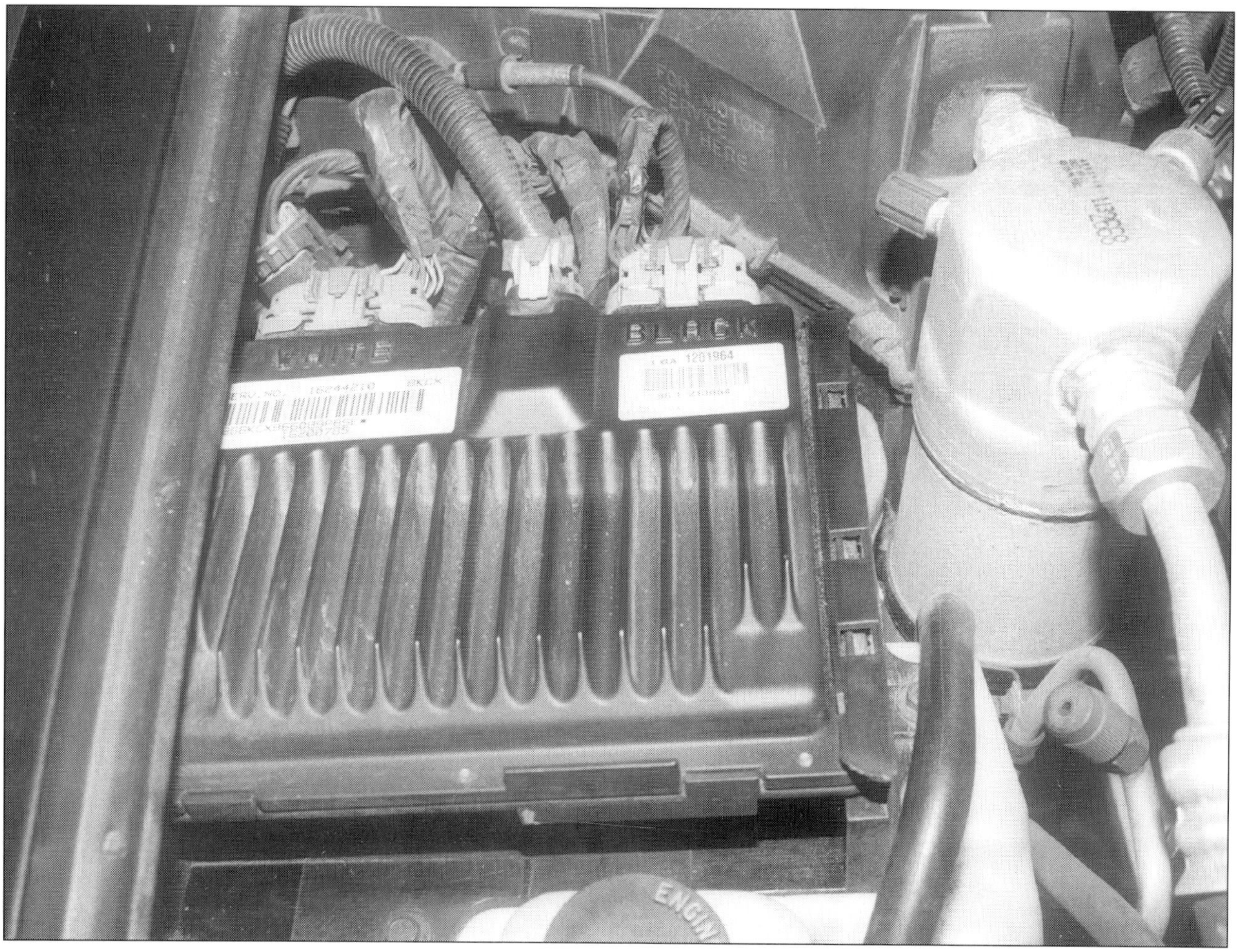

The E in "EFI" is for electronics. This is the brains of the outfit in modern GM vehicles: the PCM, or powertrain control module (aka, "the computer").

The computer relies on a myriad of sensors to determine what it should have the engine do. One of the sensors it checks is the MAT, or manifold air temperature, sensor, indicated here, relocated far forward in the intake ducting with a Howell Engine Developments MAT relocation kit. The MAT tells the computer the air temperature, and that's important because cooler air is more dense so it can hold more fuel than warmer air. Thus it produces more power. *Howell Engine Developments*

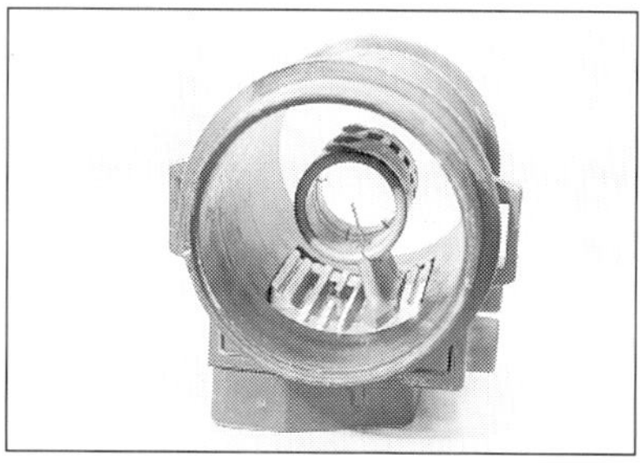

Another sensor relied upon heavily in many TPI systems is the mass airflow (MAF) sensor. This is an early-style Bosch MAF that has been modified by Howell Engine Developments and no longer uses the protective filter screens. As air flows through the MAF, it cools a thin, heated wire in the small venturi. The more air flowing through the venturi, the more voltage needs to be added to maintain a set wire temperature, so the voltage used directly correlates to the amount of air flowing in. *Howell Engine Developments*

One specific recommendation is to switch to a performance fuel filtration system that will not only reduce any flow restriction but will also provide better filtration, usually through a tighter mesh filter element.

A second modification that can be of benefit for competition use is to install a cool can to further reduce the temperature of the fuel. Aftermarket parts companies like Moroso offer cool cans, or you can make your own by simply running a length of coiled braided stainless steel fuel line through a coffee can and adding ice to the can prior to runs.

Another modification that has become standard practice for anyone modifying TPI-style engines is the act of replacing the stock fuel pressure regulator with an adjustable unit that allows you to optimize fuel pressure for your

The newer Delco-style MAF is shown as installed just ahead of the accordion-like intake hose on this 1998 Z28 Camaro. Not all TPI cars use a MAF sensor. Some cars calculate how much fuel to add by virtue of a speed-density system that works off air temperature and pressure alone and uses those values to look up how much fuel to add for given readings in a table stored in the computer's memory. There are advantages and drawbacks to both speed-density and mass airflow systems.

engine's and driving situation's requirements. Typically this will entail increasing the fuel pressure from stock settings, to better atomize the fuel as it is sprayed from the injector nozzle, and to increase the volume of fuel sprayed when the injector fires (called "on-time"). In other words, assuming an injector is "on" (spraying fuel) for a given period of time, it will deliver a given volume of fuel at a given pressure; increase the pressure and you increase the volume of fuel delivered.

Finally, depending on the displacement and power potential of your engine, it may be necessary to replace your fuel injectors with units capable of flowing more fuel (measured in pounds per hour).

Fuel System Components

Fuel Injectors

Fuel-injected engines use fuel injector units to squirt fuel into the air flowing through the induction system before it enters the cylinders. On Throttle-Body Injection systems, the injectors are positioned directly above the throttle body bores, while on TPI engines, the injectors are located in the ends of the intake manifold's ports, aimed toward the cylinder head ports.

Just like carburetor jets, fuel injectors are available in different sizes for use with different engines. Fuel injectors are typically rated by the number of pounds of fuel they flow per hour.

Injector on-time (the length of time it is instructed to spray fuel) is controlled by the ECM/PCM and depends on the amount of air, the air temperature (that is, density), and other factors.

Interestingly, while it might seem logical to replace fuel injectors when you make performance upgrades to other parts of the engine, as you would have rejetted the carburetor in the old days, the computer now takes care of that by triggering the injectors to spray fuel for a longer duration of time, if engine demands warrant it. Because the amount of time injectors spray is variable by the computer, there isn't much need to necessarily upgrade fuel injectors. In fact, upgrading injectors when they're not needed could actually kill power and cause fuel mileage to suffer because the injectors will have to fire so briefly (since each "shot" from them will contain a higher volume of fuel than would a shot from a smaller injector) that the fuel can't evenly and fully mix with the air rushing toward the cylinder. According to TPI Specialties, they have used injectors for a stock 305-inch TPI engine on engines over 400 cubic inches with no problems.

Note however, that fuel injectors do tend to respond well to higher-than-stock fuel pressures, which squirt more fuel through the injector during a given on time. Plus, they produce a finer spray for better atomization and even keep the injector itself cleaner.

Fuel Pressure Regulator

A fuel pressure regulator is used in both the Throttle Body Injection and Tuned Port Injection (including B2L/L98, LB9, LT1, LT4, and LS1) engines to maintain a constant, high pressure necessary to produce good fuel atomization when the injectors are triggered on by the ECM/PCM.

Stock fuel pressures vary depending on the engine, but are nonadjustable. Aftermarket fuel pressure regulators typically allow easy adjustment (up or down) of the fuel pressure to optimize engine performance for the given conditions.

The little black rectangular box on the intake (arrow) is a manifold air pressure (MAP) sensor on a speed-density–controlled 1993 LT1 Z28. It senses the air pressure in the intake manifold much like a vacuum gauge.

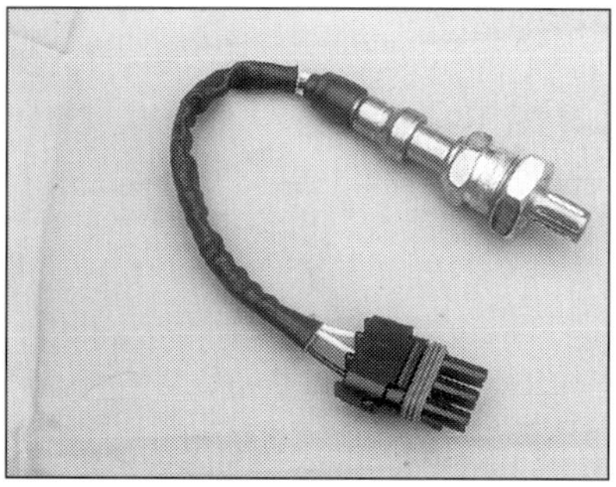

Perhaps the most important sensor, at least in terms of emissions and efficiency, is the oxygen sensor, installed in the exhaust system to check the oxygen content of the exhaust gases. If there is too much oxygen, the mixture is lean, so the O2 sensor tells the PCM to richen the mixture. If there's too little oxygen, the opposite happens. This is an ACCEL three-wire heated sensor, which works especially well in vehicles equipped with exhaust headers (instead of stock manifolds) because it reaches operating temperature much faster.

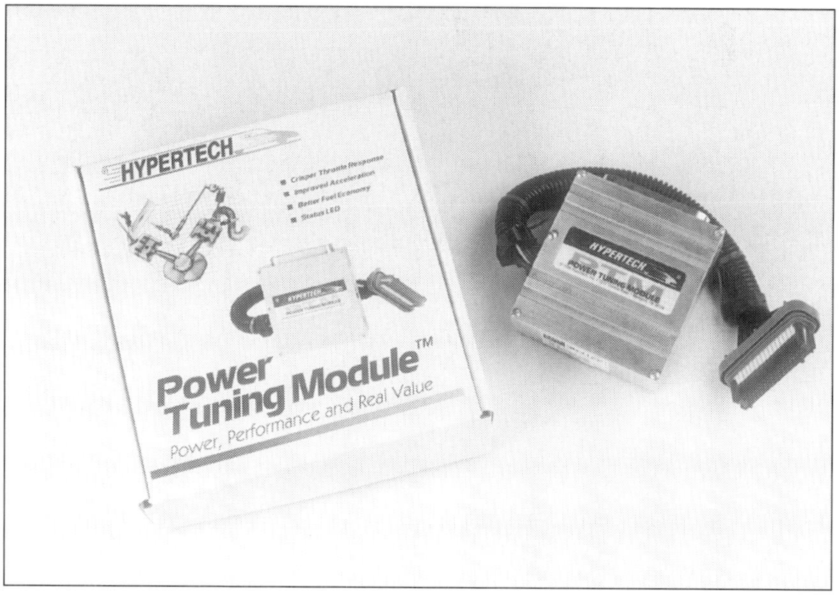

In 1994, GM changed the ECM. Not only did engineers give the computer control over the transmission (which caused a name change to PCM—powertrain control module), but the company did away with the PROM in favor of an EEPROM (electronically erasable programmable read-only memory), so the computer could be a sealed unit. To reprogram some PCMs, use of a new adapter module, like this one from HyperTech, modified the computer's processing of sensor data. *Hypertech*

There are a number of ways to alter how the computer responds to the various sensor inputs. Through 1993, the ECM (electronic control module) used a plug-in PROM (programmable read only memory) chip that held data on a number of operating parameters—what to do under what conditions. Swapping the stock PROM for "hotter" one, like this one from HyperTech, was a quick, easy way to boost power. It was sort of like rejetting the carb and recurving the distributor in one, simple plug-in module.

The performance improvements from fuel pressure adjustments are surprising and make installation of an adjustable fuel pressure regulator a highly recommended first upgrade.

Fuel Pump

TBI and TPI system fuel pumps are high-volume units, usually mounted inside the fuel tank in factory applications. They deliver an ample supply of fuel to develop 400-plus horsepower. Use of a fuel pressure regulator is required with electric fuel pumps, and installation of a fuel pressure gauge makes accurate adjustments much simpler. Fortunately, Chevrolet made it easy for enthusiasts (and service technicians) to attach fuel pressure gauges: There's a dedicated fuel pressure metering port on one of the fuel injector logs on TPI-style engines.

Fuel Filter

EFI engines use a canister-style fuel filter mounted in the fuel line between the fuel tank and the engine. The filter must be changed periodically, as it will eventually become increasingly restrictive, thanks to the debris (we hope) it traps.

Aftermarket filters that offer superior filtering abilities and reduced restriction to fuel flow are available.

Fuel Tank

The fuel tanks in most EFI vehicles are nothing particularly spectacular—they're just large metal cans that usually hold about 15 to 20 gallons of your favorite blend of gasoline. Third-generation 1LE Camaros, however, featured a special

Other PCMs, however, could take full advantage of the EEPROM by accepting uploaded programming from a small computer like this one from Hypertech. The computer first downloaded the stock PCM parameters from the PCM, then uploaded the performance parameters. Once the performance data was uploaded, the unit could then be used only to restore the original, stock parameters. This was an antipiracy feature. *Hypertech*

dual-pickup fuel pickup assembly that helped prevent fuel starvation when cornering with little fuel remaining in the

EEPROM reprogrammers and adapter modules interfaced with the PCM through the diagnostic port usually located under the instrument panel.

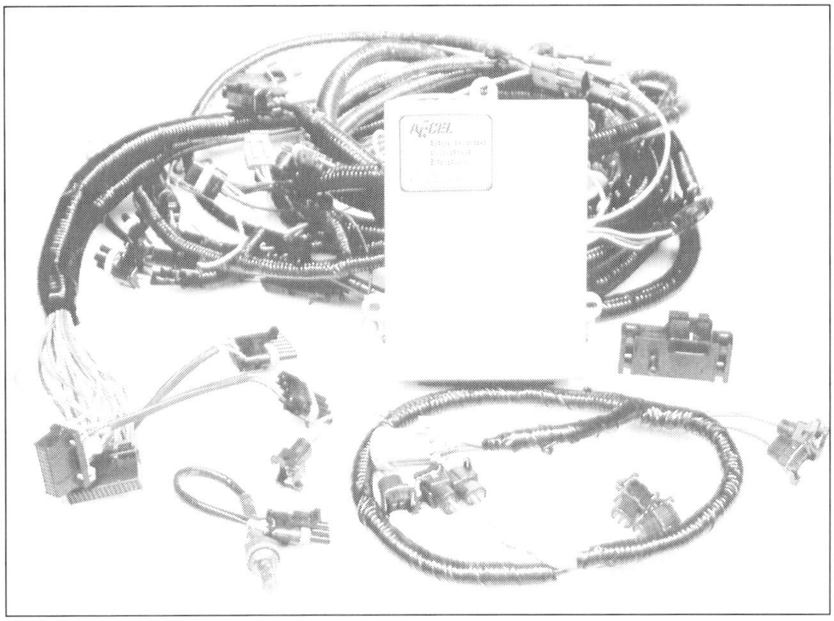

If retrofitting a TPI-style system to a non-EFI vehicle, aftermarket engine management systems like this one from ACCEL are a real blessing. Not only do they contain the wiring harness you'll need but also the computer itself and even the sensors! It's also a great upgrade for a stock system, because the ACCEL unit is easier to program. *ACCEL*

fuel tank. If you happen to road race a 1982 to 1992 Camaro, this is a worthwhile improvement, though it is of little value on the street or in drag racing or autocross competition.

Fuel

The fuel you burn in your EFI engine will determine the amount of power your engine produces, its driving character, and more. All EFI Chevrolet engines are designed to run on unleaded gasoline only. Leaded gasoline should never be used because it will plug up the catalytic converter and destroy the oxygen sensor.

Because your ECM/PCM will alter ignition timing to eliminate engine knock—which occurs much more frequently with low-octane fuels—it is wise to use premium fuels whenever maximum performance is desired, because spark ignition can be delayed until the piston is closer to top dead center and cylinder pressures are elevated. With the air and fuel charge more highly compressed, it is on the verge of self-igniting (which we refer to as "detonation") so when the spark plug fires, the gases burn quickly and produce intense heat, plus, by the time the combustion process is generating any significant downward force on the piston, the piston has already passed TDC and is traveling down on the power stroke, which minimizes any slowing of the piston's ascent on the compression stroke. High-octane unleaded racing fuels would theoretically allow the engine to operate at its peak, because all parameters could be optimized by the ECM/PCM for producing power, since engine knock would not be a concern. However, racing fuels are expensive and hard to find, making them impractical for street use.

Engine and Vehicle Management Systems

An engine or vehicle management system is the brain and nervous system of your car or truck. A computer accepts input from an array of sensors attached to the engine, transmission, and even the vehicle itself, and calculates optimal fuel mixtures, ignition timing, and even gear selections to suit driving conditions.

Production General Motors vehicle management systems for EFI engines through 1993 controlled only the engine's operations and thus were termed Engine Control Modules (ECM). In 1994, Chevrolet introduced the PCM—Powertrain Control Module—which also took over decisions for the transmission.

Later systems were termed Vehicle Control Modules (VCMs) because they are also responsible for nonpowertrain systems, such as automatic headlight systems, automatic climate control systems, antilock braking systems, traction control systems, security systems, and even equipment such as memory functions for power seat and radio station settings.

While it is possible to swap production vehicle management systems into vintage vehicles, the process can be difficult, and the systems are not easily tunable to maximize performance of modified vehicles.

Several aftermarket companies manufacture engine management systems that are designed both to ease engine swaps and to allow users to maximize their engine's performance by adjusting various parameters either through a control interface or by using software that you run on a standard personal computer.

In addition to the ECU (electronic control unit, or "the computer"), vehicle management systems use complex wiring harnesses and numerous sensors to keep tabs on various operating parameters.

Engine Management System Components

The following section identifies the major sensors and components of the engine management system and describes recommended or possible performance modifications for them.

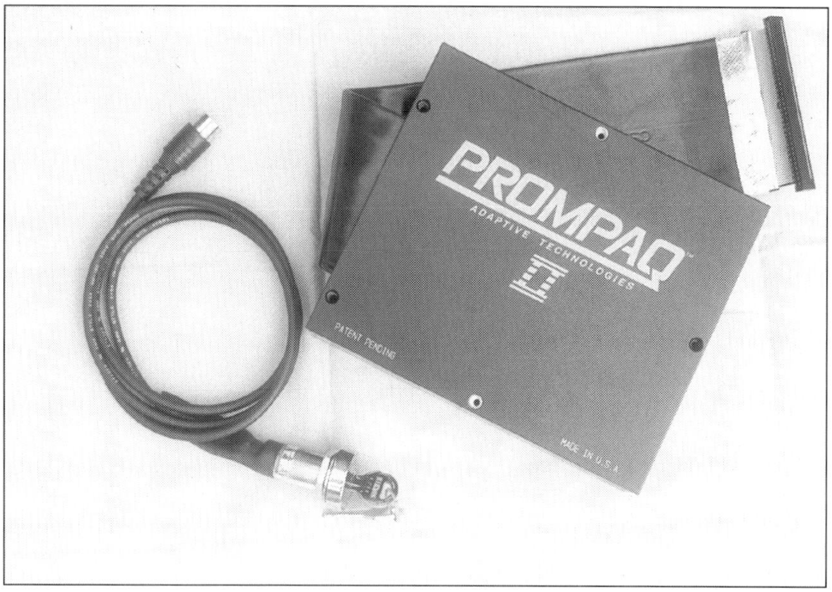

A handy add-on for PROM-equipped ECMs is Adaptive Technologies' PromPaq, which allows you to connect multiple PROM chips to your ECM and select one with the simple turn of a key.

Another handy diagnostic tool—for EFI or non-EFI vehicles—is Howell Engine Developments' Air Fuel Ratio meter, which uses an oxygen sensor to determine the air/fuel ratio, which you can then use to performance tune fuel delivery for maximum power and efficiency. *Howell Engine Developments*

Mass Airflow (MAF) Sensor

On 1985 to 1989 and 1994-and-newer TPI-type engines, a mass airflow (MAF) sensor is used to measure the amount of air flowing into the engine. That information is then fed to the engine control module (ECM) or powertrain control module (PCM), depending on the year of the vehicle, which calculates how much fuel to mix with the air to achieve a 14.7:1 air-fuel ratio. The 14.7:1 ratio typically produces the most power, best economy, and the fewest harmful emissions. The MAF sensor measures the amount of air with a thin heated wire: As air flows by the wire, it cools the wire, so the computer sends more power to the wire to bring its temperature back up to its original level. The more air flowing into the system, the more power will be required to keep the wire hot. Thus, the current is directly proportional to the amount of air flowing into the system.

The 1985 to 1989 MAF sensors used a tubular housing, and the sensor was located inside it. Screens on the housing's inlet and outlet added further protection against debris damaging the sensor (they're very, very fragile), but the screens created quite an airflow restriction: According to TPI Specialties' tests, in stock form the MAF could flow 529 cubic feet per minute, but with the screens removed, flow increases to 711 cubic feet per minute—an increase of 182 cubic feet per minute! With potential gains like that, it's no wonder that removing those screens is a popular hop-up trick, but be warned that doing so does increase your chances of ruining the MAF sensor, requiring you to purchase a costly new one. The 1992-and-newer MAF sensors are of a different design and do not require modifications, as they present only a minimal restriction to airflow. These new-style MAFs are used on Gen-II engines (except the 1993 Camaro/Firebird LT1), the Gen-III LS1 and the Vortec engines.

The 1990 to 1993 Camaro TPI engines (including the LT1) use a speed-density system to calculate the amount of air flowing into the engine. Manifold absolute pressure, manifold air temperature, and exhaust gas oxygen sensors feed data to the ECM/PCM, which uses the data to calculate how much air must be flowing into the engine. The system is much less costly, but it is less forgiving than mass air systems, because the system relies on "look-up tables"—values stored in the ECM/PCM that tell how much air is entering the system given certain MAP, MAT, and oxygen sensor values. But the stock look-up tables are based on how much air the stock system can ingest. If you change how much air the engine can ingest by installing a freer-breathing air filter, housing and cold air system, swap-in a hotter camshaft and bolt on some headers to speed the flow of air out of the engine,

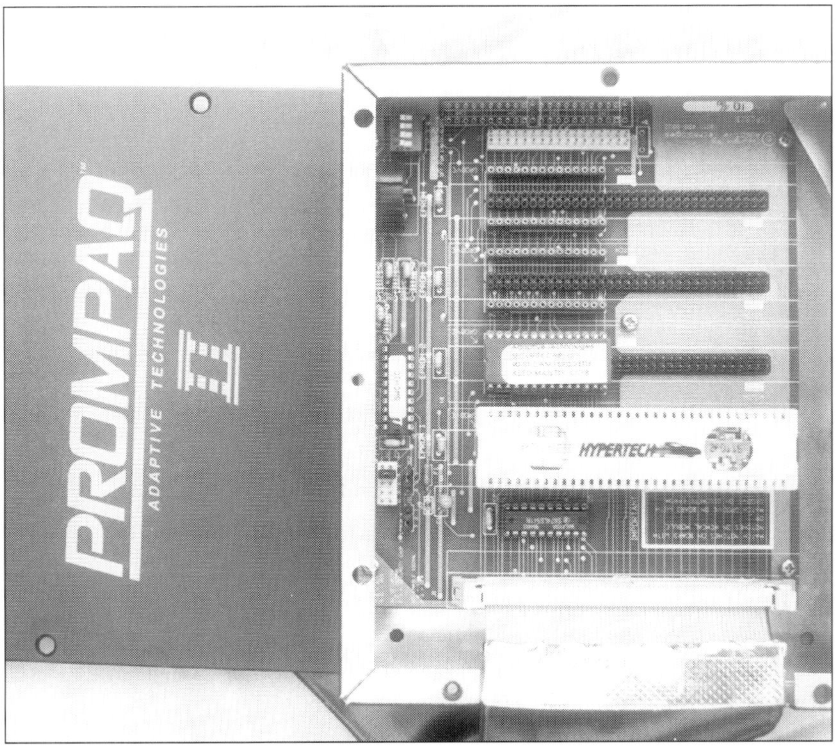

PROMs are plugged into the appropriate slots inside the PromPaq unit, and the unit is tucked away out of site (if desired). The only part that need be accessible is the key switch. There's even an available security chip that will prevent the car from starting after it is shut off, which can be handy when forced to deal with valets.

One of the most useful upgrades doesn't really do anything to your computer but rather lets you see what's going on with your engine. Rinda Technologies' Diacom software and interface cables allow you to use a regular MS-DOS/Windows PC to see the very same data being fed to your PCM, as well as the signals sent by the PCM to various motors, the fuel injectors, and other devices. It's invaluable in tuning an EFI car, especially when used with a laptop for as-you-drive sampling of data for later analysis.

the actual volume of air flowing into the engine may not be included in the stock look-up tables, so the system won't operate efficiently. In these cases, a replacement PROM chip that is custom-tailored to your engine's air intake abilities is required to restore efficiency and thus maximize power.

Interestingly, Speed-Density fuel-injection systems are more popular in professional motorsports activities that allow fuel injection, because Speed-Density systems can react more quickly to changes in rpm and engine load. But, again, the system requires look-up tables that are optimized for its particular combination of components, which can be inconvenient (at best) for the average performance enthusiast. Mass air systems, since they only measure the air flowing into the engine, are much more flexible and will still function properly whether you have a stock engine or one equipped with high-compression pistons, a huge cam, headers, and even a supercharger.

Manifold Absolute Pressure (MAP) Sensor

The manifold absolute pressure, or MAP, sensor measures the actual air pressure within the intake manifold. You might question that last statement, given that for years and years you've always talked about engine vacuum. But, the truth of the matter is that intake manifolds don't experience a vacuum; technically, they just have a pressure that is lower than the surrounding atmosphere, but is still actually a pressure (anything below 0 pressure is a vacuum).

Speed-density injection systems use a MAP sensor to determine the air pressure in the intake, then combine that data with data from the manifold absolute temperature (MAT) sensor (which reads air temperature, which tells its density) and the oxygen sensor in the exhaust system to determine how much fuel needs to be added to achieve the optimal 14.7:1 air-fuel mixture ratio.

Manifold Air Temperature (MAT) Sensor

The manifold air temperature sensor measures the temperature of the incoming air into the engine because that information tells the ECM/PCM the air's density, which directly relates to how much fuel the air will require in order to achieve a 14.7:1 air-fuel mixture for maximum performance but minimal harmful exhaust emissions.

CONTENTS

CHAPTER FIVE

CAMSHAFT AND VALVETRAIN

No other part determines as much of your engine's character as its camshaft. The camshaft determines when valves begin to open—in relation both to the crankshaft and to each other—how quickly they open, how far they open, how long they stay open, and how quickly they close. All of this dictates how much air can flow into and out of your engine, and, as we've already covered elsewhere in this book, an engine is essentially nothing more than a fancy air pump.

Of course, a camshaft alone does nothing without the various components that it controls, especially the lifters (also called tappets), rocker arms, and valves. Together, these items and others make up your engine's valvetrain and fine-tune its airflow capabilities.

Chevrolet's EFI V-6 and V-8 engines use a single camshaft mounted in the engine block, between the cylinder banks. The camshaft in a production engine is driven by a timing chain that runs around gears on the front of the crankshaft and camshaft. The camshaft spins at exactly half the speed of the crankshaft, thus creating the four strokes of modern internal combustion engines: intake, compression, power, and exhaust.

The camshaft has egg-shaped lobes that raise lifters that ride on the lobes, in turn raising the pushrods. Each pushrod raises one end of a stud-mounted rocker arm, causing the other end to push downward against the tip the valve. As the lifter crests the peak of the cam lobe, a spring connected to the valve by way of a retainer and keepers closes the valve and keeps tension in the system to prevent parts from falling out all over the place.

All Chevrolet small-blocks since the first one in 1955 have used this same basic valvetrain system, though with some changes along the way for improved performance and durability. Most notably was the change in 1987 to a roller-tappet camshaft, and the use of roller rocker arms on the 1996 LT4 engine. But there have been smaller changes too, like the switch to rail-type rocker arms in 1988, or the factory's redesigned springs for the LT1 in 1992. As much as the small-block Chevy is heralded for its parts interchangeability,

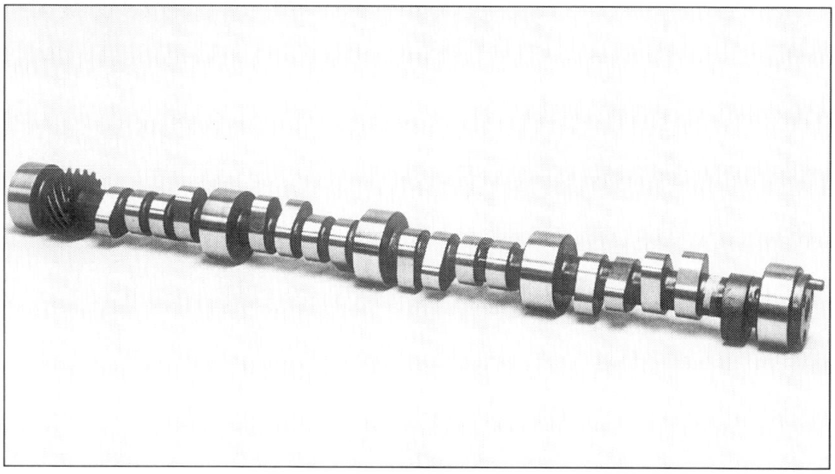

Camshafts are the key to engine performance. The cam dictates when, how far, and how long each valve opens, and that controls how much air can get into or out of the engine. Despite being a single piece, they're extremely complex. That makes picking the right cam a difficult task. *TPI Specialties*

There are basically two types of camshafts: flat tappet (left) and roller. Because of the small wheel on the end of a roller lifter, the lifter can follow a much more aggressive lobe profile, so valves can open much more quickly to their peak and stay at that peak longer, thus allowing more air to flow past the valve. Additionally, each type of cam is available in a mechanical (aka, "solid") or hydraulic version. Mechanical cams tend to work better at high rpm, but hydraulics require less maintenance and run quieter.

changes such as these make it a risky proposition to swap parts—especially on EFI engines, when the majority of these changes were being implemented.

Camshafts

Because your camshaft determines how and when your engine's valves open, it determines how much air moves in and out of the engine (intake, exhaust, and head flow capabilities notwithstanding). So, changing your camshaft for a larger grind can radically alter the performance characteristics of the engine.

Myron Cottrell of TPI Specialties recommends that customers swap camshafts first, actually, rather than make a series of small changes with bolt-on parts that combined still won't yield the gains of the cam swap. Short of nitrous oxide or forced induction systems, swapping the camshaft is the quickest way to get the most noticeable power gains.

But choosing a camshaft isn't exactly child's play. For starters, there are two basic types of camshafts, each with two different types of lifters, or lobe followers. Beyond that, you've got the actual characteristics of the cam lobes to consider, such as valve lift and duration, plus the intake and closing timing events—of each of the two valves. And then there are the lobe centerlines, the lobe separation angles (and overlap) and more, each of which will distinctly affect how your engine will perform.

Fortunately, unless you're trying to design your own custom-ground camshaft, you won't have to deal with all of this. Camshaft manufacturers have already explored the possibilities and are more than happy to help you select a camshaft that will work well in your particular vehicle, with the powertrain components with which you've equipped it.

A quick rundown, however, will tell you that unless you're running a pre-1987 block, you'll be running a roller camshaft, because that's all 1988-and-newer blocks will accommodate. Pre-1987 blocks can use either flat-tappet cams or roller cams (though only with aftermarket roller lifters), but there's really no advantage to using a flat-tappet cam.

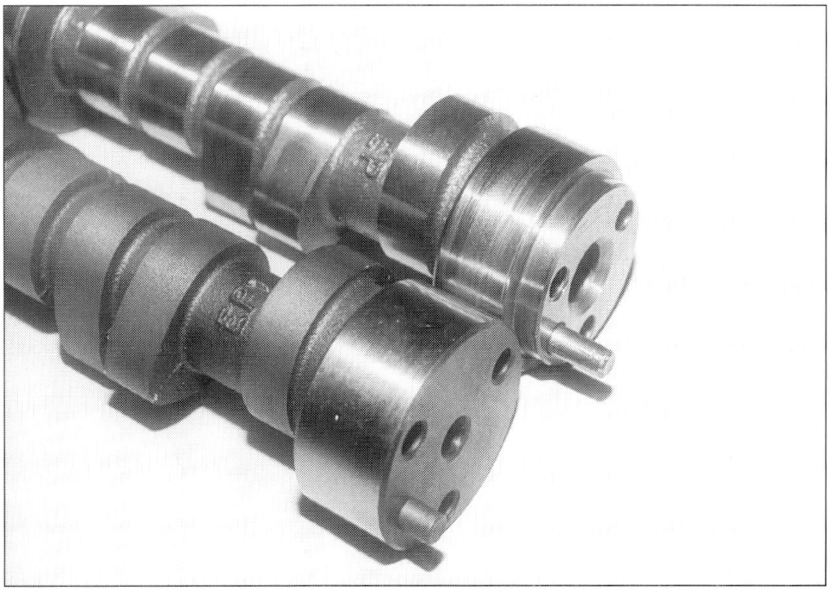

While interchangeability is a hallmark of Chevy's small-block, when Chevy changed to roller cams in 1988, it changed more than just the lobe shape. Note the "stepped" cam snout.

The Gen-III LS1 cam resembles first- and second-generation small-block cams, but they actually have a hollow core for lighter weight, a larger base circle for added strength, and larger bearing journals for reduced wear. All Gen-III cams are roller cams.

Roller camshafts offer significant advantages in terms of valve opening and closing rates. Flat-tappet cams have to have gradual camshaft lobe opening and closing ramps in order to allow the tappet to follow the lobe profile without digging into it, thus with flat tappet cams, any given valve spends the bulk of its time either opening or closing, but spends very little time (measured in degrees of crankshaft rotation) open at its peak point of valve lift. Roller lifters, though, by virtue of the large rolling wheel on the their base, can follow a much steeper ramp, so during the same number of degrees of crank rotation (that is, "time") the valve wastes fewer degrees opening and closing and instead spends the bulk of its time open at maximum lift, posing less of a restriction to airflow by the valve, so the cylinders fill and empty more easily. Another way of thinking of this, is to picture a triangle and a square, each with equal length bases and the same height. Flat tappet cam lobes are shaped somewhat like a triangle, whereas roller cam lobes are more like squares (they're not square, but for the simplicity of discussion, we'll assume they are). Relying on grade school geometry lessons—or even just visualizing the triangle superimposed in front of the square—it's easy to see that the square presents much more area than the triangle, which correlates to the amount of area beneath the valve when it is open.

Put in English, picture this utterly preposterous scenario: You're a participant in a wacky money-grabbing contest that requires you to start on the ground floor of a high-rise building, race up to the 50th floor where the money is, grab as much of the cash as you can (there are plenty of bags to put it in) in a given allotted time (say two minutes), then race back to the ground floor. You have the option of either climbing the stairs to the 50th floor or taking the elevator. If you choose to climb the stairs (assuming you could even make it to the 50th floor in one minute) you would immediately have to turn around and run back down the stairs in order to get back to the ground floor before your two minutes is up, so you would only have a second or two to grab a little money. This would be like the flat tappet cam, because the valve spends all its time opening (climbing) and closing (descending). On the other hand, suppose that you can reach the 50th floor within 30 seconds by taking the elevator, and that the return trip took the same 30 seconds. Thus, you would spend only half your time (one minute) getting to the 50th floor and back, so you would have a whole minute to grab a ton of cash. Obviously, by taking the elevator you'll wind up with more money than by taking the stairs, so that makes more sense. Roller cams work in a similar sort of way: they allow the air and fuel charge more time to flow into the cylinders. Aftermarket roller cams are available for pre-1987 engines originally equipped

At the rear of the Gen-III cam, just ahead of the rear bearing journal, is a cam position indicator. The computer compares data from the cam position sensor with that from the crank sensor to determine not only what position the crankshaft is in, but what stroke the engine is on, since the crank will be at any one position twice in one four-stroke cycle. This allows more precise ignition and fuel injector timing, among other things.

Camshaft Terminology

Base Circle: The portion of the camshaft lobe during which no lift is generated, after the intake valve has closed but before the exhaust valve begins to open.
Duration: How long, in crankshaft degrees, the valve remains open. Note that there are generally two figures quoted: "advertised" duration and duration at 0.050-inch. The "at 0.050" figure is a standard for comparing two cams, whereas advertised figures are not consistently measured from one cam to the next or from one manufacturer to the next.
Lash: The clearance between the rocker arm and valve stem tip, in inches, required by mechanical tappet (flat or roller) camshafts to allow for heat expansion. Lash figures are usually different for intake valves than for exhaust valves and are provided on a camshaft specification card provided with new camshafts.
Lift: Expresses how high, in inches, a cam lobe rises above the cam's base circle (lobe lift) or how high the valve is raised from the valve seat (valve lift). Note that valve lift is the lobe lift multiplied by the rocker arm ratio. For example, a cam with 0.300-inch lobe lift and a 1.5:1 rocker produces 0.450-inch valve lift. With a 1.6:1 rocker it produces 0.480-inch valve lift.
Lobe Centerline: The point, in crankshaft degrees, of the exact center (usually peak) of the cam lobe, either intake or exhaust.
Lobe Separation: The number of degrees of crankshaft rotation between the centerlines of the intake and exhaust valves of a given cylinder.
Overlap: The number of degrees of crankshaft rotation during which both the exhaust and intake valves of a given cylinder are both open (as the former is closing and the latter is opening). Can be calculated by adding the intake valve's opening point to the exhaust valve's closing point.
Ramps: The portion of the cam lobe during which the valve is either opening or closing.
Single Pattern/Dual Pattern: Single-pattern camshafts have the same lift and duration for both the intake and exhaust lobes. Dual-pattern cams have differing lift or duration specifications for the intake and exhaust lobes, usually to allow the exhaust valve to remain open longer to increase exhaust flow out of the cylinder to compensate for weaker cylinder head exhaust ports.
Valve Timing: The points, as measured in crankshaft rotational degrees, when the valves either open or close.

with flat-tappet cams, but do require use of aftermarket roller lifters.

Note that while it is possible to compare lift and lobe separation angles between flat-tappet and roller cams, you cannot directly compare duration figures—even duration at 0.050-inch figures—because roller cams will tolerate much more aggressive opening and closing ramps, which causes their duration figures to be considerably higher. Put another way, a roller cam can reach a specific opening point (say, the industry standard of measurement: 0.050-inch of valve lift) sooner, and, likewise, the valve can close more quickly from 0.050-inch more quickly, so we've effectively increased duration of the time (in degrees) the valve is open at both the beginning and end of the valve event. So while the measurements may be taken at the same dimension of valve lift, the roller cam figure is considerably larger, so the roller cam will behave entirely different—and that doesn't necessarily mean better.

One of these LS1 cams is the stock one, and one of them is Callaway Cars' super-hot SuperNatural cam. Based on the visible lobe profiles, which do you think is which?

Factory Cams

Through the years, Chevrolet has used a number of different camshafts for its EFI engines, so many, in fact that it would be impossible (and pointless) to discuss each one.

It is helpful to understand the general patterns of stock cams, however, and how they fall short for performance use.

Nearly all GM cams favor low-speed torque, a smooth idle, and clean emissions over all else. For most vehicles and most drivers, that's a wise decision, of course, but for enthusiasts who want more performance, power tapers off

Cams should also be degreed to ensure that they are properly phased with the crankshaft. The process is a lot easier than many people tend to believe. And it's the only way to ensure accurate adjustment of a cam, should you choose to alter its phasing by advancing or retarding it a few degrees. This is Competition Cams' hyper-accurate large-diameter degree wheel. *Competition Cams*

When installing camshafts, always make sure to apply the manufacturer's recommended cam lube. The thick, grease-like goop prevents wear in the cam's most critical time of need: its first 20 minutes of life. Cams also need plenty of oil in those first minutes, so engines with new cams should be run at 2,000 rpm for the 20 minutes immediately after starting it.

quickly with these cams at even modest rpm, limiting their effectiveness in competition or even spirited street driving.

Valve lift is generally kept well below 0.500 inch. In fact, with the exception of the Impala SS's cast-iron version of the LT1 head and the pre-1999 Vortec V-8 heads, most iron small-block Chevy cylinder heads won't flow well beyond 0.420-inch of valve lift. (For more on cylinder heads, see chapter 3.)

Long duration figures give the cylinder more time to fill at higher rpm, but sacrifice low-end power, so Chevy typically keeps roller cam duration specs under 220 degrees of crankshaft rotation. Likewise, tighter lobe separation creates more valve overlap, which not only bleeds off cylinder pressure but hurts idle quality and dilutes the intake charge with exhaust gas that actually flows back into the intake. Lobe separation figures of 114 degrees or more are the norm on factory cams and rarely go below 112 degrees.

General Camshaft Recommendations

Making camshaft recommendations for unspecified engine combinations is like throwing darts at a moving target—when you're blindfolded. While there are some general specs that will *usually* work well, ultimately the only way to determine whether a camshaft will work well in your engine is to try it. Fortunately, aside from the several hours you'll spend performing the swap, the investment isn't outrageously expensive.

With that in mind, here are some recommendations that have to be taken with a very large grain of salt.

High-Performance Street Driving

If you really want to have fun on the street, you want your engine to make gobs of power below 4,000 rpm. For that, you'll want a cam that features shorter duration figures because they'll trap cylinder pressure at lower rpm, when there's actually plenty of time for the cylinders to fill. Wider lobe separation angles are desirable for the same reason: They don't bleed-off cylinder pressure and they don't dilute the fresh air/fuel mixture with exhaust gases. Another figure that is often better is valve lift: The farther you lift the valve from its seat, the more air can flow past the valve to either enter or exit the cylinder.

A good upgrade for the LT1 or an L98 engine is Chevy's own LT4 cam. It features a 218-degree duration (at 0.050-inch tappet lift) for the intake and 228 degrees on the exhaust side. These figures are significantly shorter than many aftermarket cams, favoring low-speed torque, but with 1.6:1 rocker arms (as in the LT4) the cam develops 0.525-inch lift at the valves (both intake and exhaust; 0.492 inch with a 1.5:1 rocker) which helps preserve decent high-rpm breathing. Lobe

Many racers choose a belt-drive timing set, like this Jesel system. The belt runs more smoothly and is actually more accurate, plus it allows easy cam adjustments. Gear drive systems are also available, but tend to transmit undesirable harmonics to the cam, which can cause spark scatter on engines with rear-mounted distributors.

separation of 112 degrees again favors low- to midrange power because little cylinder pressure is bled off and the intake charge remains more pure than it would be if lobe separation were tighter.

For comparison, early L98 350 TPI roller cams featured just 202/206 degrees of duration at 0.050-inch lift—too short for good power—and only slightly more than 0.400 inch of valve lift (0.403-inch intake and 0.415-inch exhaust) with a much broader lobe separation of 115 degrees. With this L98 cam, duration is so short that the cylinder doesn't get much of a chance to fill completely before the intake valve shuts; lift is minimal too, so the engine shuts off early; the 115-degree lobe, however, traps lots of cylinder pressure, which gives the engine a smooth idle and good throttle response.

If, by chance, you happen to be working on a big-inch stroker motor, you're in luck. Thanks to the torque advantages larger displacement engines enjoy, these engines are less sensitive to changes that rob low-rpm power. For such engines, narrower lobe separations and longer duration often provide results worth smiling over.

Drag Racing

Low-end response isn't of much concern on a drag strip, so tighter lobe separations and longer durations can be selected to promote upper-rpm cylinder filling. Valve lifts, likewise, are generally much greater than on the street, as well, which requires careful attention to valve-to-piston clearance.

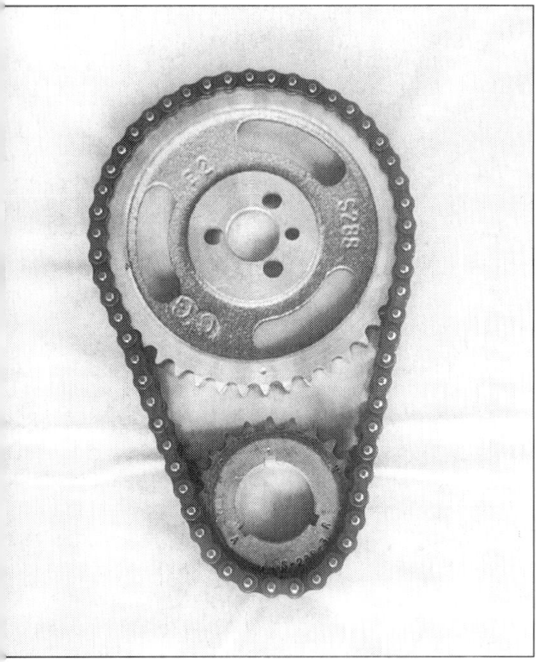

By far the most popular method of harnessing the crankshaft's rotation to turn the cam is with a chain-and-gear combination. For high-performance engines, a double-row, true roller timing chain set is far more preferable than a less expensive link-belt type set.

On first-generation engines, it's wise to install a cam button spacer to prevent the cam from "walking" forward in the block, which leads to erratic valve actions. Buttons are made in aluminum, plastic, or in a metal "roller" version; be aware that the aluminum buttons can wear against the inside of the timing cover, causing metal fragments to end up in the oil.

Recommendations are much more difficult to pin down on competition engines because there are just so many possibilities of other equipment and modifications that may have been made to an engine, not to mention the differences between vehicles into which the engine may be installed.

One general improvement for drag racing that can be made is to try retarding the camshaft a few degrees, which will usually boost top-end power. The process for retarding (or advancing) the cam is pretty straight-forward: You enlarge the index hole on the cam's sprocket and install an adapter that relocates the index hole. Different adapters provide differing degrees of relocation.

Autocrossing

Autocrossing will typically involve a production-based vehicle, which usually means a Corvette, Camaro, or Firebird. Under these assumptions, it's possible to make a few more assumptions about what sorts of changes will be beneficial.

Tight, twisting courses demand crisp throttle response and low-speed acceleration. A stock cam can actually work well for autocross competition, especially if installed slightly advanced, to open the intake valves a bit earlier than usual.

If you feel your stock cam isn't sufficient—even after trying it advanced by 2 or possibly 4 degrees—a replacement cam may be your best solution. One with relatively short duration but high lift will typically result in the characteristics needed for autocrossing: crisp throttle response and plenty of low-end power. Similarly, lobe separation should be modest, in the 112 to 114 degrees ballpark, to prevent reduced cylinder pressures.

Road Racing

Depending on class and sanctioning body rules, you may not have much choice of what cam you run: Most classes specify a stock cam, or at least one that tests stock. If you find yourself in this situation, your only choice may be: Do you want to advance the cam 2 degrees or 4?

Of course, cams that test stock aren't always stock. TPI Specialties' Super Profile L98 cam is similar enough to a stock cam that it will fit within the rules, but it features significantly extended duration figures to keep the valves open longer: 218 degrees for the intake and 219 for the exhaust, compared to 207 and 213, respectively. These changes, coupled with the cam's 108-degree lobe separation keep your engine pulling strong at upper rpm, well beyond where the stock cam would have fallen flat. Valve lift, however, remains stock modest at 0.415 inch for the intake and 0.433 for the exhaust side.

Assuming you're not encumbered by such rules, and you've got the equipment to take advantage of it, you can go for a cam with higher lift and even longer duration. Production LT1 and pre-1999 Vortec V-8 heads will generally work well with valve lift of 0.500 inch or so, but much more than that and you'll need better heads for a proper match. Increased duration—to as much as 240 degrees—will greatly enhance upper rpm breathing; however, don't expect such cams to idle smoothly or be good for commuting to work.

The Valvetrain

A camshaft is just a big lumpy stick of iron (or steel, in the case of roller cams). Alone it doesn't do anything except take up space in the block. It needs an entire cast of supporting characters to even allow your engine to run.

Unfortunately, the parts it calls upon are considered to be among the weakest parts in your engine. Often it is not the reciprocating assembly of your engine that limits rpm, but rather the complex valvetrain assembly. So, as enthusiasts start making more power and looking for more speed (which ultimately results in increased engine speeds), there is a distinct need for more durable valvetrain components. The fact that aftermarket performance components also tend to provide other benefits is merely icing on the cake.

Understanding Your Valvetrain

The first thing to understand about the valvetrain is that its job is to convert the round-and-round motion of the camshaft into an up-and-down motion at the valves. Unfortunately, the antiquated design of the Gen-I and Gen-II small-blocks relies on complicated valvetrain geometry that impairs reliability, plus imposes certain limitations on valvetrain upgrades. Specifically, if viewed from above, the lifter is not located in the same plane as the valve tip, so the pushrod is at a slight angle, which means the rocker doesn't merely pivot, but

Gimme a Lift?

One of the things to be extremely careful of is changing valve lift. While valve lift is often beneficial, more than any other single cam spec change it can get you into trouble quickly, if you open the valves so far that they contact the piston. If that happens, you can bet the valve, piston, head, and probably cylinder walls are all going to wind up destroyed. While excessive cam lift is often the culprit, even a relatively modest cam can run into trouble if installed in an engine that has been rebuilt with higher compression pistons, shaved deck surfaces (head or block), or otherwise-reduced piston-to-deck clearance. The general rule of thumb is to ensure you have at least 0.100 inch of clearance at top dead center. You can check this by placing a layer of modeling clay on the piston during assembly, installing the head and valvetrain assembly, and rotating the crankshaft two complete revolutions (one complete cam revolution). Then remove the head and measure the thickness of the clay (section it with a sharp knife).

Flat tappet lifters, as these Competition Cams lifters show, are aptly named. *Competition Cams*

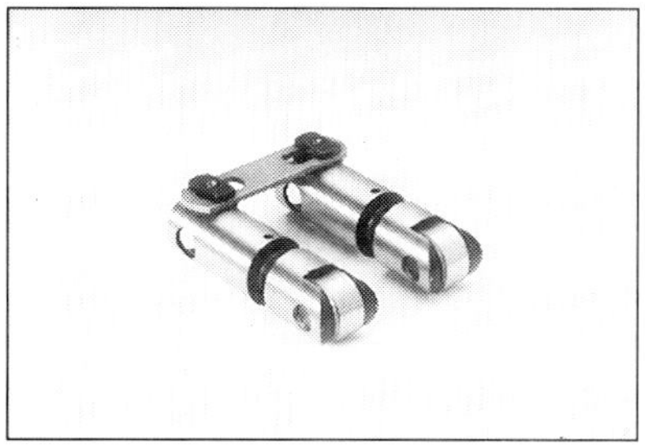

Roller lifters, likewise, are so named for equally obvious reasons. *Competition Cams*

rather endures some side-loading. In addition, if viewed from the front of the engine, the pushrod base remains stationary, but its top does not, because the arc of the rocker arm pushrod cup causes the rocker tip to sweep inward then outward, which can cause the pushrod to bend. Furthermore, the tip of the rocker arm also swings through an arc, so it slides across the tip of the valve, and actually changes the leverage ratio in the process.

The second thing to understand is that your first instinct to simply replace everything with heavy-duty parts is wrong. Sure, heavy-duty pushrods, oversize springs, and other parts will make the valvetrain more reliable, but those parts will also increase the weight of the valvetrain reducing the engine speed at which valve float will occur. Essentially, the more mass your valvetrain has, the harder it is for the valve springs to reverse the direction of the valve's travel. When the kinetic energy, or movement, of the valvetrain components is so great that the valve spring can no longer cause the valve to begin closing when the lifter crests the peak of the cam lobe, the valve stays floating open in the cylinder longer than it should. If you're lucky, the worst that will happen to your engine when this occurs is that your valve springs will overheat, causing them to lose tension, reducing the speed at which valve float will occur next time. If you're unlucky (and who of us isn't?), the valve may contact the piston, causing the valve head to bend and maybe destroying the piston and combustion chamber in the process. Or worse (yes, it can get worse), the valve spring retainer and keepers could fail to keep their grip on the valve. If that happens, the valve will drop into the cylinder, irreparably destroying not only that cylinder's piston, cylinder walls, and combustion chamber, but also causing damage throughout the engine as little bits of debris ricochet back into the intake and make their ways to other cylinders.

Planning Valvetrain Upgrades

So the goal of any modifications to your valvetrain should be to increase reliability without necessarily increasing the mass of the components—and preferably decreasing the mass, which would extend the upper rpm limit.

Another problem that valvetrains encounter is flex, usually at the rocker stud. Minimizing flex at the rocker stud keeps valvetrain geometry more precise, minimizing the chance of problems while also ensuring that you get full and proper movement of the valves.

Speaking of movement of the valves, rocker arms are designed to amplify the cam lobe's profile, specifically the lift. Stock rockers (except the LT4, LS1, and 1999+ Vortec V-8s) use 1.5:1 rocker ratios, which means that the valve end of the rocker travels 1.5 times as far as the pushrod end of the rocker. At least, that is what production rockers are supposed to do. Unfortunately, production stamped steel rockers are not

Note the offset pushrod cup in the Competition Cams lifter in the foreground. Such "trick" solutions are really only necessary on all-out race engines in order to correct for poor valvetrain geometry or to compensate for revised cylinder head port locations. *Competition Cams*

highly accurate, which causes different cylinders to have different valve lift and duration actions, which naturally affects cylinder-to-cylinder power balance. In addition to not being accurate, stamped steel rockers aren't efficient either because they drag their tip across the valve stem tip (which creates friction) plus they scrape against a pivot "ball" at the rocker stud (more friction). So, in one fell swoop you can eliminate rocker ratio variations to improve cylinder-to-cylinder balance, and at the same time decrease friction and reduce valvetrain mass, simply by installing aftermarket aluminum roller rockers. Going a step farther, you can increase the ratio of the rockers (as Chevy did with the LT4, LS1, and new 1999+ Vortec V-8s) by switching to rockers with a 1.6:1 ratio, which will cause the valve to lift farther from the valve seat.

The remaining possibilities involve upgrading the valves and their various related hardware. Because valves have so much to do with airflow through the cylinder head, they are discussed more thoroughly in chapter 3; however, suffice it to say that production iron valves are neither the lightest nor the strongest valves on the market.

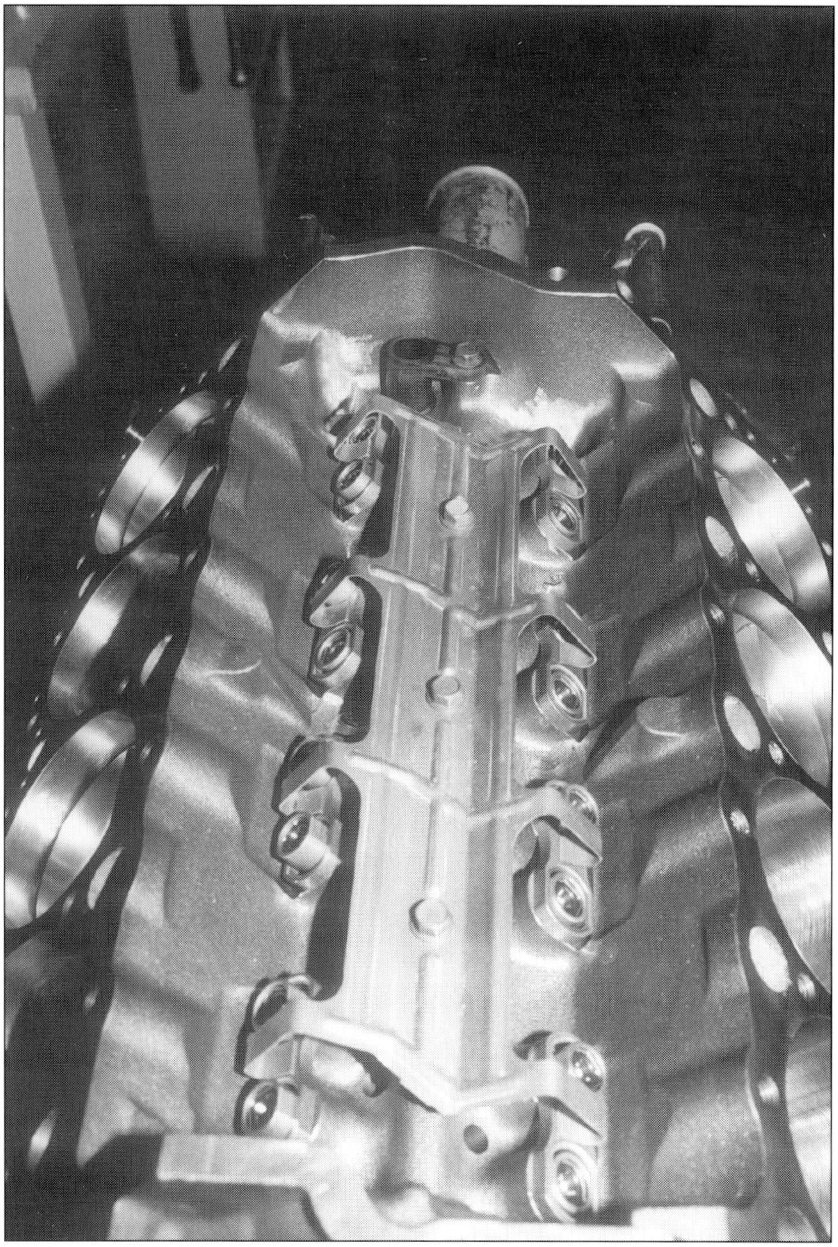

Factory roller lifters are held in place by this sheet metal spider that bolts into the lifter valley.

Production valve springs (again, except the LT4, LS1, and 1999+ Vortec V-8s) are single-coil springs with 1.24-inch outside diameter and a flat-wire internal damper. Production V-8 valve springs typically featured an 80-pound seat pressure at an installed height of 1.7 inches. Until the 1992 LT1 debuted, however, the springs featured a rate of about 270 pounds per inch. But aggressive roller cams were challenging the spring, so for the LT1, Chevrolet upped the rate to about 370 pounds per inch, but managed to keep the seat pressure roughly the same as the old spring. This spring works well with a variety of camshaft profiles; however, more aggressive profiles will exceed this spring's ability to control the valve too.

If you end up changing valve springs, you may need to change valve spring retainers too. Even if your new springs don't require them, though, you may still want to consider replacing the retainers because stronger and lighter replacements are available. Another aspect of spring retainers that some enthusiasts believe is important to consider upgrading is the angle of the valve keeper pocket in the retainer. Production retainers feature a 7-degree pocket and matching pair of keepers, while flatter 10-degree keepers and retainers are available as performance upgrades.

Two final concerns are the valve seals and the valve guides. The seals are designed to control the amount of oil that can travel down the valve guide into the valve pocket area of the cylinder head. As we covered more fully in chapter 3, valve guides are simply a replaceable sleeve in which the valve travels up and down. Because the valve is mounted at an angle (23 degrees in production Gen-I and Gen-II heads, 15 degrees in LS1 and 1999+ Vortec V-8s), some wear of the valve guide is inevitable, eventually, since the rocker actually causes the valve stem to rock back and forth (toward and away from the center of the engine), which pounds out the valve guide and creates excessive play in the guide. By replacing the valve guides—or resleeving your existing guides—you can restore proper valve geometry so it aligns properly with the seat. (Valve guides are covered more thoroughly in chapter 3.)

Valvetrain Components

The following section describes the basic functions of the various valvetrain components, the challenges those components face, and potential upgrades for different situations.

Cam Drive (Timing) Sets

The first part a cam needs is a drive system—something to turn it. Camshafts are driven by the crankshaft. A sprocket on the cam and another on the crank reduce the speed of the cam to half that of the crankshaft. Chains, gears, or ribbed belts are used to connect the sprockets, and each has its particular advantages and disadvantages.

Chain-drive systems are inexpensive and relatively quiet; however, the chain is subject to stretch and that causes erratic timing. Double-row true roller chain-drive sets minimize this possibility, while reducing friction at the same time to free up power.

Converting the pushrod's upward motion into a downward force that opens the valve is the teeter-totter-like rocker arm, like these stock, stamped-steel units. Stock rockers are fine for stock engines, but for high-performance engines there are better choices.

Gear-drive systems are much stronger than chains, which eventually stretch; however, because gear-drive systems so accurately transfer the crankshaft's rotation to the cam, they can cause unwanted harmonics to be transmitted too. Most gear drives are noisy, though more recent designs are quieter, and thus less objectionable for street use. Gear drives can employ either a single gear between the cam and crank gears or twin gears, and the systems may or may not fit beneath stock timing chain covers.

Belt-drive systems are very quiet, very accurate, easily adjustable, and feature an easily replaceable belt for when the belt eventually stretches. The catch is that the belts do stretch and thus need to be replaced rather frequently. Belts also have to be kept dry; if they come in contact with oil they will slip, which can cause the valves to actually come in contact with

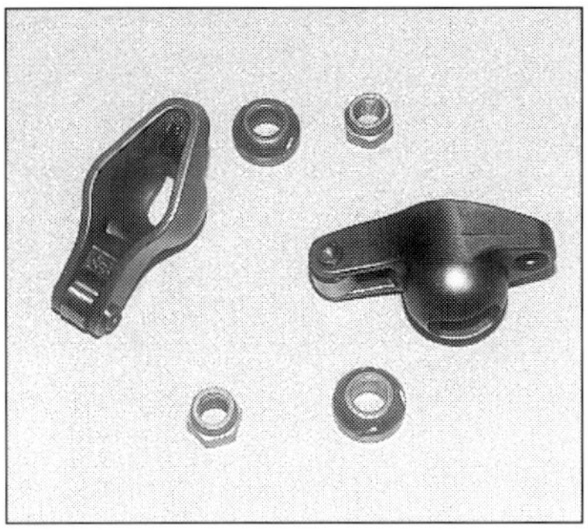

Aftermarket rockers come in many types, from more accurate versions of stock-type stamped steel units to ones like these Competition Cams Magnum roller-tip rockers that are stronger than stock rockers and generate a little less friction.

the pistons—a definite no-no. Furthermore, belt drive systems are somewhat wider than chain and gear systems, which can make them difficult to use in some circumstances, since they won't fit under standard timing covers.

Because the driven gear mounts to the camshaft with bolts, it is wise to install the bolts with a bolt lock-plate, which features bendable ears that prevent the bolts from loosening over time. Such lock-plates are very inexpensive (less than $5) and provide excellent insurance against what could be a serious problem.

Also note that Chevrolet has changed the cam snouts a few times, requiring different cam gears for each version. Make sure your cam gear is the proper match for your cam.

Tappets/Lifters

Tappets—or lifters, as they're often called—ride along the cam lobe. As the camshaft spins, the lobe's shape causes the tappet to rise up and down in its bore, which in turn raises the pushrod, thus tipping the rocker arm which pushes the valve open. (And the leg bone's connected to the hip bone. . . .)

Four types of tappets are available: mechanical flat-tappet, hydraulic flat-tappet, mechanical roller, and hydraulic roller. Mechanical tappets (flat or roller) generally will sustain higher engine rpm before floating off the cam's lobe, which makes them popular choices for racing engines. Mechanical lifters require a slight clearance between the rocker arm nose and the tip of the valve (to allow for heat expansion), however, and this clearance causes a ticking noise during engine operation as the rocker strikes the valve tip. The vibrations and shock caused by the ticking also tends to loosen the rocker

Competition Cams' Hi-Tech Stainless rocker arms feature not only a roller tip, but a roller fulcrum plus a super-strong stainless steel alloy to minimize deflection and increase durability. They're also rebuildable, so a set can theoretically last forever. *Competition Cams*

arm, requiring periodic adjustment to maintain proper valve operation.

Hydraulic tappets work somewhat like shock absorbers—the oil within them cushions the movements of the valvetrain, so hydraulic lifters run much quieter. But because the tappet's pushrod seat takes time to reseat after the lifter crests the peak of the cam lobe, hydraulic lifters will lose contact with the cam lobe several hundred rpm earlier than will mechanical lifters. Because of the cushioning effect hydraulic lifters provide, they do not induce harsh vibrations into the valvetrain, as do mechanical lifters, so lash settings are maintained and thus hydraulic lifters require far less frequent adjustment, and usually extend the life of other valvetrain components, such as pushrods and valve stems.

Depending on your penchant for performing routine maintenance, use of solid or mechanical lifters allow for higher operating rpm than do hydraulic lifters, which tend to pump up much above 5,000 rpm. This shouldn't be confused with valve float (which is related to the valve springs being unable to do their job of returning the valve to its seat). Lifter pump-up actually refers to the fact that, at high engine speeds, the oil gets squirted out of the lifter faster than fresh oil can flow back in, so the lifters effectively end up pumped-up with air, which results in a lot of slack in the valvetrain, and thus unpredictable valvetrain action. Note that the particular style of tappets, or lifters, that your engine requires is dictated by the camshaft you install, since each camshaft is designed to work only with a particular style of lifter.

Degreeing a Cam

Most enthusiasts have heard of degreeing a cam, which is simply the process of precisely determining when the valves start opening in relation to the crankshaft's position. The process isn't difficult, nor does it take much time to complete.

The primary reason to degree a cam is that most cams and cam drive systems are mass-produced, which means they are manufactured with slightly varying acceptable tolerances that could result in your camshaft not providing the precise valve event timing it was designed for (the proper specifications are listed on the cam card provided with new cams). Even a 1-degree change can result in a dramatic power loss in a high-performance engine.

In addition, by degreeing your cam you can precisely advance or retard the cam's relationship to the crank position to alter the engine's power characteristics. Advancing the cam a couple of degrees typically livens up low-end power, but at the expense of top-end abilities. Retarding the cam has the opposite effect: better top-end but less downstairs at streetable engine speeds. Note that even though many aftermarket timing gears have key slots for 2-degree advanced and retarded positions, manufacturing tolerances may allow those keys to be out of position by as much as a degree.

If, while degreeing your cam with a degree wheel, piston stop, and pointer, you discover the cam is mistimed, you can correct that flaw by installing indexing inserts in the cam gear bolt holes.

Also note that old flat-tappet lifters should never be used with a new camshaft because the lifters will already have developed a wear pattern that may cause accelerated wear of the cam lobe. Likewise, if you ever remove old lifters and intend to reinstall them with the same camshaft, make sure that you reinstall each lifter in its original position, since lifters and lobes wear together to form compatible mating surfaces. No such problem exists when using new lifters with an old cam, nor are roller lifters subject to these rules.

Pushrods

Pushrods have two functions: First, they are responsible for transmitting the lift provided by the tappet up to the rocker arm. Second, a hollow channel inside the pushrod acts as a passage to deliver oil to the rocker arm and valve assembly at the top of the cylinder head. One end of the pushrod rests in a cup in the tappet, and the other sits in a similar cup in the rocker arm. Holes in the lifter and rocker arm allow oil to enter and exit the pushrod.

Unfortunately, pushrods suffer from a bad reputation that isn't entirely deserved. As pushrods attempt to transmit the lift from the cam lobe to the rocker arm, they are subjected to some complex, compound geometry that is necessitated by the need to avoid intake ports and water passages in the head while traveling through an arc dictated by the rocker arm. This geometry causes the pushrod to experience some sideloading, which can lead to bending. The forces become greater at higher rpm and with higher valve spring pressures. Still, pushrods—especially hardened performance units—are generally able to withstand the forces that try to bend them. Big monkey wrenches are thrown into the works, however, if the engine experiences valve float, because valve float introduces slack into the system, which allows the already complex valvetrain geometry to become even more complicated as the rockers pivot sideways and even begin to spin on their studs, increasing the side loading of the pushrods, causing them to bend.

Factory pushrods are 5/16 inch in diameter and made of steel. They are reasonably strong, but are not generally able to withstand sustained operation over 5,500 rpm without risk of bending. Aftermarket pushrods are available in both larger diameters (3/8 inch is a common upgrade) and stronger materials, such as chrome-moly steel. Despite the added weight, large diameter, stronger pushrods are a worthy investment, especially for engines that will see sustained high rpm; they will increase engine reliability and provide added insurance against failure.

Aluminum roller rockers, like these Crane aluminum gold racing rocker arms, offer light weight, high strength, and greatly reduced friction.

Depending on the type of rocker arms and rocker studs being used, you may also need to run pushrod guide plates, to minimize pushrod side-to-side movement. With the exception of one company's plastic parts, pushrod guide plates are made of steel and require hardened pushrods to prevent wear when the pushrod rubs against the guide plate.

Rocker Arms

Rocker arms are responsible for converting the pushrod's lift into a downward pushing force that will lift the valve off of its seat. They pivot on a rocker arm stud that is either pressed or screwed into the cylinder head (more information on rocker studs follows this section).

Production rocker arms on all Chevrolet V-8 engines (except the LT4, LS1, and 1999+ Vortec V-8 engines) are of stamped steel construction and do not feature rollerized tips or fulcrums. Beginning in 1988, Chevrolet began using self-guiding rocker arms, which feature a slight trough on the tip of the rocker, with ridges on each side that straddle the valve stem tip. Self-guiding and old-style nonguided rockers are not interchangeable, despite their similar appearances. In addition, self-guiding rockers require slightly longer valve stem tips, to prevent contact between the rocker tip and the spring retainer.

The 1996 LT4 engine uses aftermarket 1.6:1 forged aluminum, fully rollerized rockers, while the Gen-III engines use exceptionally strong but lightweight powdered-metal 1.6:1 rockers with roller fulcrums and tips.

The standard rocker arm ratio is 1.5:1, meaning that a camshaft lobe lift of 0.300-inch is amplified one and a half times by the rocker arm, resulting in 0.450-inch of valve lift. However, because of production tolerances, production rocker ratios actually vary considerably. It's not uncommon for an engine to be equipped with one or more rockers that actually measure 1.48:1 and others that may be as high as 1.52:1 or more. While that may not sound like much, a 1.48:1 rocker ratio yields 0.444 inch of lift from a 0.300-inch lobe, while a 1.52:1 ratio will yield 0.456 inch of lift from the same lobe—a difference of 0.012 inch. An inconsistency like that can greatly affect overall engine efficiency and causes

Good Things Come in Small Packages

Unlike picking a camshaft for a non-emissions-equipped, carbureted Chevy engine, buying a cam for a TPI engine is a little less chaotic, if only because you have fewer cams from which to pick. But that doesn't mean you can afford to be any less careful. For one thing, cams for TPI engines tend to cost between three to five times more than a non-TPI cam.

The best line of advice to follow when picking a camshaft is to be conservative. Pick a cam that provides moderate lift increases, moderate increases in duration, and moderate changes in lobe separation. Huge cams rarely perform well in any conditions other than sustained high-rpm use. Small cams, however, which "pop" the valves open quickly, then pop them back closed in similar fashion and have relatively wide lobe separation, will generally yield better throttle response, and therefore a better "seat of the pants" feel.

unbalanced cylinder-to-cylinder power output.

Aftermarket rockers are typically much more exact in their measurements; however, a 1.5:1 ratio aftermarket rocker arm often opens the valve slightly more than one and a half times the lobe lift, to compensate for the rocker's arc, which decreases the rocker ratio. As the rocker tip swings downward, it is also dragged back across the valve stem tip, which wastes some of the pushrod's lift, resulting in a reduced effective ratio. In addition, because the portion of the rocker that contacts the valve stem tip is slightly curved (to prevent it from digging into the valve tip as it slides across it), the ratio at zero lift is ever so slightly greater than the ratio at peak lift. Aftermarket manufacturers combat both of these flaws by lengthening the rocker tip, thereby increasing the effective ratio, often to around 1.52:1.

Nonroller stamped steel rockers are available from aftermarket performance parts companies, but roller versions are preferable, especially rockers with roller fulcrums (pivots) which greatly reduce friction and can result in a 10-degree or better reduction in oil temperatures, not to mention their effect on freeing up power. Fully rollerized aftermarket rockers are typically constructed of either cast or forged aluminum, while roller-tipped rockers are typically steel.

Because rocker arms multiply the lobe lift, using a rocker arm with a higher ratio—such as 1.6:1 or even 1.7:1—is essentially like swapping the camshaft for a bigger cam. For instance, a 0.300-inch lobe lift translates to 0.450 inch of valve lift with a 1.5:1 ratio, 0.480 inch of lift with a 1.6:1 ratio, and 0.510 inch with 1.7:1 rockers. Thus, an hour spent changing the rocker arms could boost valve lift by more than .060 inch! It's interesting to note that some camshafts and cylinder heads respond well to a mixed rocker ratio setup, such as 1.5:1 rockers for the intake valves, but 1.6:1 rockers for the exhaust valves, to allow the cylinders to empty out the exhaust more completely, so as not to contaminate the fresh intake charge, which would obviously decrease power output. Consult your camshaft and cylinder head manufacturers or machinists for their rocker ratio recommendations.

Rocker Arm Studs and Girdles

Rocker studs have the seemingly simple job of locating the rocker arm in a fixed position. Stock rocker studs are 3/8 inch in diameter and are either pressed into the cylinder heads or screwed in. All aluminum cylinder heads use screw-in rocker studs, and screw-in studs are always preferable, since pressed-in studs can loosen and actually pull out of the head. Heads that have pressed-in studs can be machined to receive screw-in studs. Screw-in studs do require pushrod guide plates, which typically require hardened pushrods. Use of guide plates on a cylinder head not originally equipped with them may also require that the holes in the cylinder head through which the pushrods travel be slightly enlarged for added clearance.

Studs are available in a larger 7/16 inch diameter, which will resist bending and flexing better than the thinner 3/8 inch units, and thus result in more accurate valvetrain action. And since the portion of the studs that thread into the head are the same 7/16-inch size for both the 3/8-inch and 7/16-inch rocker studs, swapping to the beefier studs is a simple task. Note, however, that larger diameter rocker studs do require rockers designed for the larger stud diameter.

For racing applications that will see extremely high rpm, it may be useful to use a rocker stud girdle, which is essentially just two long pieces of metal (usually aluminum) that are bolted together, one on each side of the studs, to tie each stud to all the others on that head for additional strength and resistance to flexing.

Valve Seals

Valve seals are used to limit the amount of oil that is used to lubricate the valve stem and valve guide. Two basic types are used: the production-style O-rings and the performance-oriented PC, or umbrella, seals that clamp around the head of the valve guide.

PC-style guides are preferable to the stock O-rings, though PC seals may require modifications to the valve guides for proper installation and clearance between the seal and the spring retainer.

O-rings deserve a quick installation note too. Novice engine builders often unwittingly install them incorrectly: They must be installed *after* the spring retainer, while the spring is compressed. Then the keepers can be installed and the spring can be decompressed.

Valve Springs

While the camshaft is responsible for opening the valve, the valve spring has the duty of closing the valve, once the lifter/tappet crests the peak of the cam lobe. Valve springs must be matched to the camshaft to ensure that the springs are not so strong that they actually cause the lifter/tappet to wear the cam lobe.

Valve springs weaken with age and mileage, and new springs should always be used when installing a new camshaft. Get the springs that the cam manufacturer recommends.

When installing valve springs, always ensure that you install them at the specified installed height, using shims, if necessary, to increase the height of the spring plus any shims, or having the head's spring seats machined if the

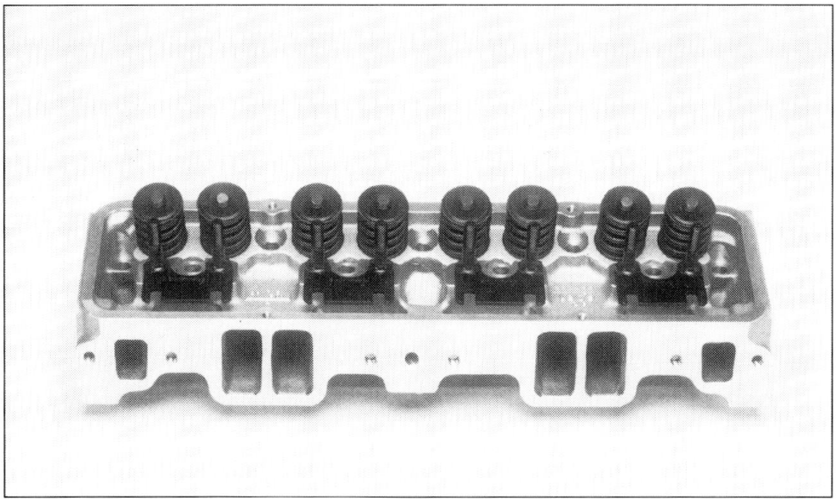

This aluminum Edelbrock Performer cylinder head is equipped with pushrod guide plates (in front of the springs) to keep the tip of the rocker aligned with the tip of the valve. Beginning in 1988, GM began using self-aligning rocker arms that feature a slight groove in the tip to keep them on the valve tip, so guide plates are not required. *Edelbrock*

Valve springs are deceptively simple. They have the job of returning the spring to its seat in a controlled manner, once the lifter has crested the peak of the cam lobe; however, a number of things can create problems, including weak springs that allow valve "float"; coil bind when compressed too far; and interference problems with rocker arms, head bolts, and more.

spring's nonshimmed installation height is taller than specified. Often, however, no two springs will be precisely the same, nor will the spring pockets in the head (or spring cups, for aluminum heads). So before any costly machine work, try swapping the spring to another position and remeasuring them.

It's also important to inspect for coil binding. With high-lift cams or thick-wire springs, it's possible for two or more coils of the spring to physically contact each other when the spring is compressed. This situation will quickly bend pushrods and has been known to cause rockers to break and even to suck valve keepers through the spring retainer.

With large-diameter springs and roller rockers, it's especially important to check for interference between the rocker arm and spring.

Valve Spring Retainers and Keepers

Valve spring retainers are metal discs that do exactly as their name implies—they keep the valve springs in place. The valve runs through a hole in the center of the retainer.

Production valve spring retainers are made of steel and are typically adequate for general high-performance and mild-racing use.

Aftermarket retainers are often made of higher strength or lighter weight materials, and they are available for oversize springs.

Because valve spring retainers are retained by angled valve keepers, the angle of the bore in which the keepers rest must match that of the keepers. Production keepers use a 7-degree angle, while racing keepers often feature a 10-degree angle to better prevent the keepers from pulling through the retainer at high rpm or with exceptionally strong spring pressures.

One critical inspection to make is to verify that there is sufficient clearance between the bottom of the spring retainer and the top of the valve guide and valve seal when the cam is at peak lift. In general, a 0.100-inch clearance will be ample.

CHAPTER SIX

EXHAUST SYSTEM

An efficient exhaust system is essential for a performance engine, because an engine can't draw in new air and fuel unless the old, burned exhaust gases are removed from the cylinders first. Stock exhaust systems and what kind of system you can assemble for your TPI engine will vary depending on the actual vehicle in which the engine was installed. For instance, a Corvette's exhaust system is different from a Camaro's, which in turn is different from a truck's set of pipes.

Stock Systems—Understanding the "Standard"

Corvettes have featured fairly efficient tubular steel "shorty" exhaust headers since the TPI engine first debuted. The units provide decent flow and fit into the tight confines of Corvette engine compartments nicely, without interfering with accessories or other hardware.

Camaros and Firebirds, to save on production costs, featured cast-iron log manifolds until the introduction of the LS1 motor to the F-car platform in 1998, when they at long last were granted tubular steel manifolds as well.

The C/K pickups, likewise, feature cast-iron manifolds because of their durability and sound-absorption properties.

Exiting the manifolds, exhaust gases travel through steel pipes to the catalytic converter (or multiple converters, depending on the vehicle and options). Catalytic converters attempt to clean up emissions by converting them into harmless elements. Steel pipes carry the purified gases aft to the muffler, then on to a steel tailpipe. Most late-model production exhaust systems use stainless steel exhaust pipes, which extends the life of the system by preventing rust.

As with most things on production cars, the stock GM-designed exhaust systems work reasonably well, but aren't designed for peak performance. Unfortunately, there are few modifications you

Most people think this is how you start improving your exhaust system: bolt on a set of headers. Though that's not actually a bad start, there is someplace else that should be considered first.

Real improvements need to start here, with the cylinder head's exhaust valves and ports. Cylinder heads are so complex they need their own chapter, however, so you'll find more information about them in chapter 3.

can make to an exhaust system legally, given the Environmental Protection Agency's hard-line stance against any modifications that may alter or defeat the operation of emissions control devices. So the only viable options available are to replace components with EPA-approved replacement parts.

Exhaust System Improvements

The actual exhaust system improvements you can make may be somewhat dictated by the vehicle in which your TPI engine will be installed. While custom-built street rods may feature plenty of wide-open space for installing headers and a free-flowing dual-exhaust system, other cars may not be so accommodating. Likewise, you need to consider the legality of any changes you intend to make.

While custom exhaust systems may be a possibility for your application, the reality is that most enthusiasts will be limited by the availability of ready-to-bolt-on systems. Fortunately, there are a lot of options, thanks to aftermarket companies.

The goal, of course, when trying to improve your exhaust system is to increase exhaust flow out of the engine. In the information that follows, we'll examine each (typical) exhaust system component, run through the problems that the components present, and discuss how to solve those problems.

Cylinder Head

Lots of people believe that their exhaust systems start with their exhaust manifolds or headers. They don't. The exhaust system actually starts inside your engine, in each cylinder.

For the sake of simplicity, we'll say the first part of your exhaust system is the cylinder head, though a case could easily be made that your pistons and block are also parts of the exhaust system, because of the fact that as each piston travels back up the cylinder (following the power stroke) the piston actually pushes some of the burned gases out of the cylinder and into the cylinder head.

The two critical parts of your cylinder head that affect exhaust flow are the exhaust valve and the exhaust port, which are both covered in-depth elsewhere in the book (Chapter 5 and Chapter 3, respectively). It's important to remember that your heads are part of the exhaust system, because they represent the first possible obstruction to exhaust flow, and if your exhaust system starts out choked off, any changes you make downstream of the heads can't possibly produce maximum flow.

Exhaust Manifolds/Headers

Once the exhaust leaves the head, it enters the exhaust manifold or header.

Most production cars (except 1985 and newer Corvettes and LS1-equipped cars) use cast-iron log exhaust manifolds

Stock log-type cast-iron exhaust manifolds like these on a TBI-equipped 305 truck engine barely let even a stock engine breathe enough to stay running. They're also quite heavy, which isn't exactly beneficial in terms of either handling or acceleration. However, cast-iron manifolds are quiet and they will last just about forever.

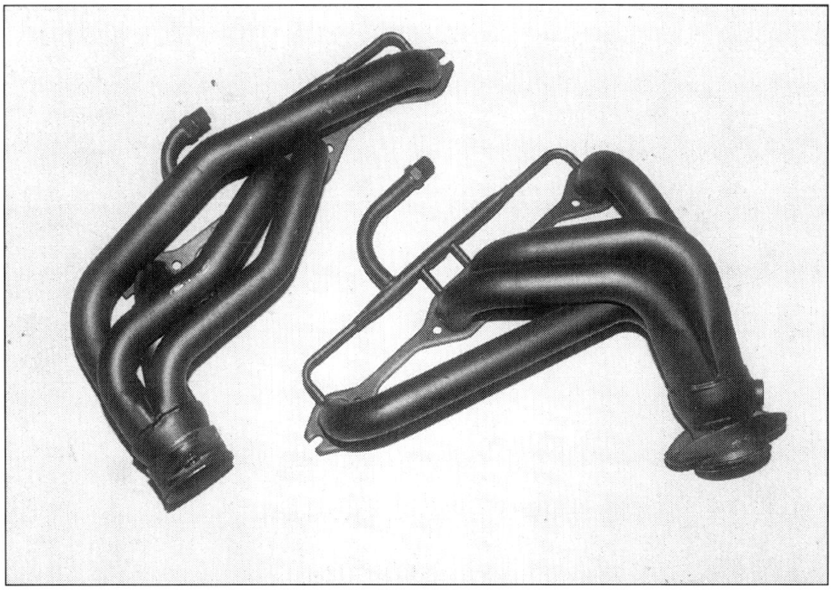

From a performance standpoint, headers are a much better route. These Edelbrock Tubular Exhaust System (TES) "shorty" headers give each exhaust port its own pipe to breath into, which eliminates most of the bottleneck of iron manifolds.

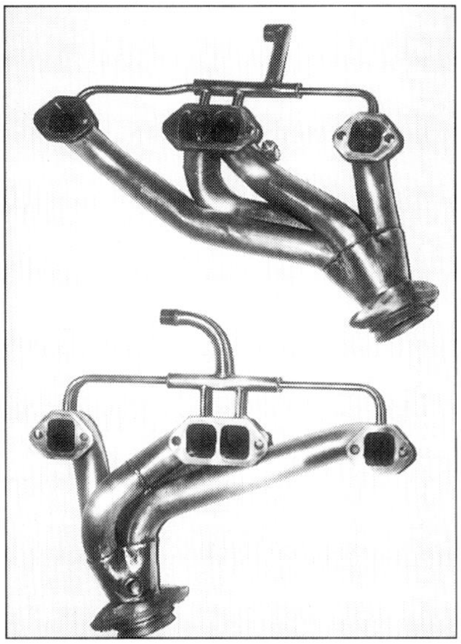

These shorty headers are from SLP and feature stainless steel tubing, which will allow them to last virtually forever, since they are highly rust-resistant. *SLP*

to direct exhaust gases into the exhaust pipes. Cast-iron exhaust manifolds are super quiet and virtually last forever, but they're also horribly restrictive and terribly heavy, neither of which makes them particularly suited to performance use. Weight-wise, the problem is obvious: They're made of cast iron—lots of cast iron. In terms of flow restriction, the problem is that all four cylinders on one side of the engine feed into one common plenum. But if that plenum is already full of exhaust from, say, cylinder number one, then the gases from cylinder number three have nowhere to go. Most of cylinder three's exhaust gas will eventually get out, but some won't. Some will be trapped in the cylinder when the exhaust valve closes, and that means that less fresh air and fuel can be drawn into the cylinder, so the engine won't make maximum power.

Aftermarket tubular exhaust headers, on the other hand, typically flow much better, since each cylinder has its own primary header tube, thus pulses of exhaust gases from individual cylinders don't have to fight for space in the single runner of the stock manifold. Long, equal-length primary tubes tend to work better than shorter (that is, shorty) tubes of unequal length, because the longer equal-length tubes deliver each exhaust pulse to the collector tube in the same order as the engine's firing order, so pulses of exhaust gases don't even have to fight for space in the collector. In fact, by preserving the firing order, each exhaust pulse that reaches the collector helps to scavenge the next pulse from the next cylinder (in the firing order on that side of the engine) by creating a low-pressure zone behind it, which the next pulses rush to fill.

That said, though, long-tube, equal-length headers won't fit into all vehicles, whereas shorty headers generally will. If given the choice between shorty headers or stock manifolds, the shorty headers will always provide a measurable power boost.

And since headers are made of relatively thin-wall steel tubing, they weigh less than iron manifolds, which both reduces the amount of power needed to move the vehicle plus helps vehicle handling. Unfortunately, the two areas headers haven't typically excelled in are durability and noise—the areas in which cast-iron manifolds really shine. The noise emitted from headers generally isn't considered objectionable (not by performance enthusiasts, anyway), however, nor is it overly loud. Furthermore, advances in header coatings have allowed headers to last virtually indefinitely without rusting, corroding, or discoloring.

Metallic-ceramic coatings do more than just make headers (or other exhaust system components) look good and last longer, though. The coatings help the header retain heat, which keeps the exhaust pulses hotter for longer. And since hot air is less dense than cool air, a given volume of it can flow more quickly, so coated headers actually do a better job of expelling exhaust gases than do noncoated headers.

Header wraps and blankets do the same thing as header coatings, in terms of retaining heat, only the wraps and blankets do it better. Unfortunately, header wraps and blankets do their job too well: They retain so much heat that the header tubes get so hot that the metal undergoes a structural change that makes it brittle enough to literally crumble like a cookie after a relatively short time. For racing applications this isn't necessarily a problem, especially if you can justify replacing headers every season or so. In that case, the benefits of the wraps and blankets far outweigh the detriments. On the street, however, the benefits provided by header wraps and blankets aren't needed and their side effects are a nuisance.

Oxygen Sensors

In closed-loop operation mode, an oxygen sensor mounted in the exhaust man-

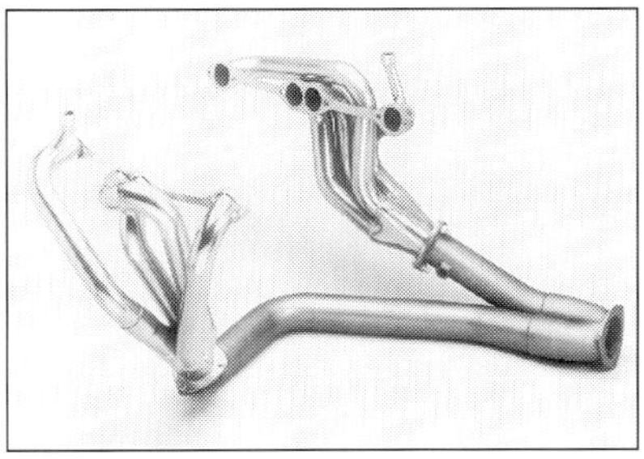

An alternative to expensive stainless steel headers are headers that feature a ceramic coating, like these from Arizona Speed and Marine. The coating both protects the pipes from rust and other forms of corrosion, but also acts as an insulator, trapping exhaust heat inside the pipe, which increases flow through the pipe for more power. *Arizona Speed and Marine*

ifold or header measures the amount of oxygen in the exhaust gases to determine how efficiently the engine is burning the air/fuel mixture. The oxygen sensor's data is used by the ECM/PCM to control the air/fuel mixture; if there is a lot of oxygen in the exhaust gases, the system is running lean, so the ECM/PCM will add more fuel by lengthening injector on time. Conversely, if the amount of oxygen in the exhaust gases is low, the system is running rich, so the ECM/PCM will lean-out the air/fuel mixture.

There are no repairs that can be made to an oxygen sensor other than replacement. The only modification of sorts is to replace a nonheated oxygen sensor with a heated (three-wire) unit that will become operational much quicker following cold starts (unheated sensors rely on exhaust gas temperature to heat them). This is especially important for vehicles equipped with headers, since the exhaust gases may cool considerably by the time they reach the oxygen sensor, thereby requiring more time to heat up to true operating temp.

In the case of cars equipped with headers, an oxygen sensor bung (fitting) can be installed in the collector pipe if the headers do not feature one from the factory.

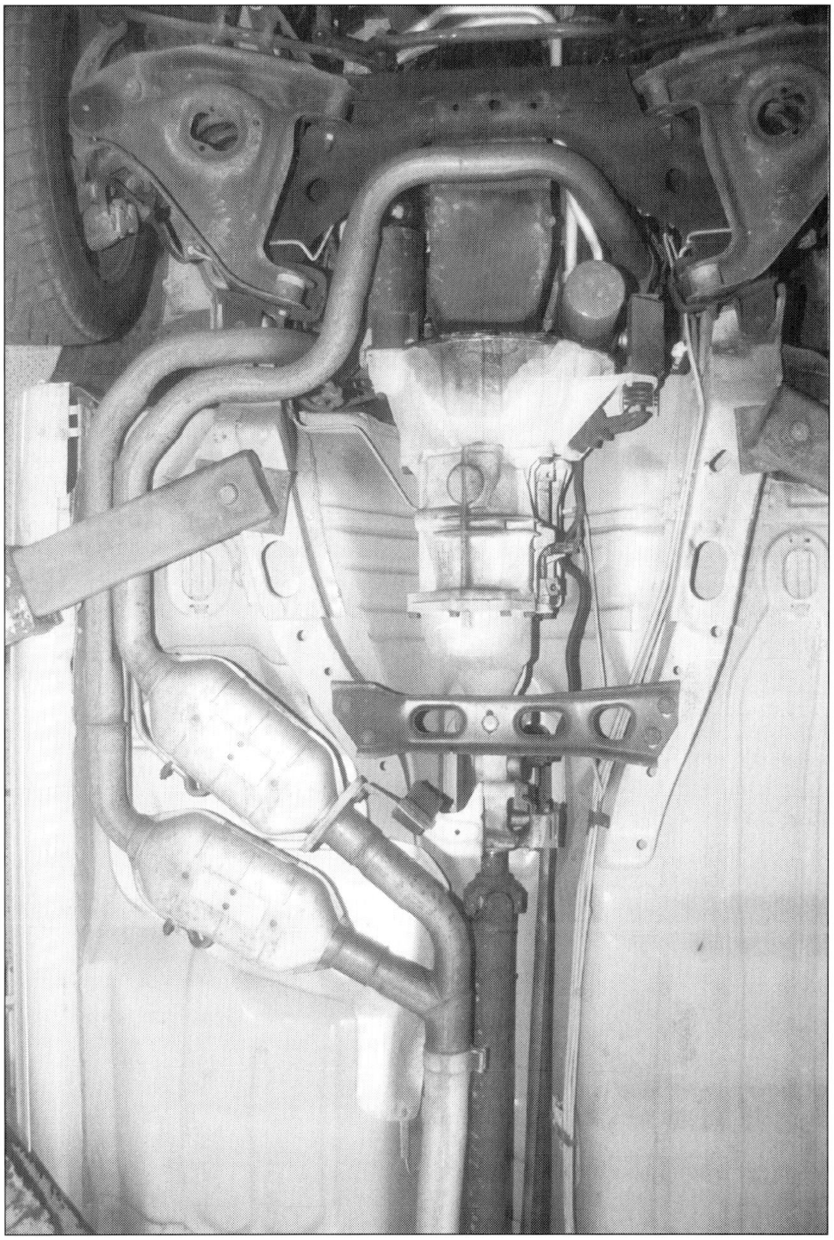

From the manifolds or headers, exhaust gases flow into the head pipe, or pipes, as this optional dual-catalyst system shows. Most vehicles feature a single catalyst to purify exhaust gases coming from each cylinder head. Unfortunately, single catalysts often aren't capable of flowing enough air to avoid being a restriction, so exhaust gases back-up in the head pipe, limiting power. Dual catalysts, however, alleviated that problem and result in a significant power increase. But due to emissions laws, they aren't necessarily available for use on every vehicle.

Primary Exhaust (Header) Pipes

Not to be confused with the pipes that make up a set of exhaust headers, the primary exhaust pipes—often called header pipes—connect the exhaust manifolds or headers to the catalytic converter. In most cases there is little that can be changed about your header pipes, since their size and shape are often dictated by the vehicle in which the engine is being installed.

Catalytic Converters

Catalytic converters are responsible for cleaning up the exhaust emissions by literally converting the harmful gases into less toxic gases thanks to a chemical reaction and the injection of oxygen into the exhaust system, which mixes with the deadly carbon monoxide and converts it to carbon dioxide, which plants and trees need to breath.

Ceramic Coated Headers

Tubular exhaust headers provide some awesome performance benefits that have made them extremely popular over the years. Specifically, headers can dramatically increase torque and horsepower, plus they're lightweight, and thereby reduce the amount of power necessary to move the vehicle. But, until recently, headers had one serious drawback that gave many enthusiasts reason to pause and debate installing headers on their vehicles: corrosion.

Since headers are generally made of mild-gauge steel tubing, they rust incredibly easily. Add to that the fact that paints can't generally stand up to the heat created by the exhaust, and most headers were forced to face the world completely unprotected from the elements. This made headers literally an unattractive choice for show cars, because, short of never driving the car, there was no way to keep them looking good. And for performance cars, they would typically rust-through after only a few years worth of use.

That all changed when high-tech ceramic coatings started hitting the market. The coatings applied a ceramic-based substance to the exterior of the headers and often to the interior as well. The substances add a durable, attractive finish that is virtually impenetrable to contaminants like water, air, and more. In addition, the substances act as a heat barrier, trapping exhaust heat in the header, so the exhaust gases remain hotter and thus less dense, and therefore able to flow more quickly through the exhaust system. Plus, by trapping heat within the header, coated headers reduce underhood temperatures, which can contribute to lower intake air temperatures, which will contribute to even more power.

For those enthusiasts who are interested in building show cars, ceramic header coatings are also available in a variety of colors, to match or contrast with a vehicle's paint scheme.

It should be noted that not all headers are coated in the same manner. In particular, some companies only spray-apply the ceramic coating to the exterior of a part, so a header's interior is still left exposed, unprotected, and susceptible to the ravages of time and the elements.

It's interesting to note that the ceramic coatings typically applied to parts like headers are actually quite flexible, so they are less likely to crack, flake, and peel off as the header pipe swells and contracts with the effects of heat. In fact, the coatings can even be applied to suspension components, like springs, and they need not be applied right after a part is manufactured. In fact, several companies offer coating services, which work especially well for custom-built headers or for customization of parts for show cars.

Naturally, the perfect complement to a set of coated headers would be a coated exhaust system, but often the cost of such a system is prohibitive. Stainless steel pipes are a good substitute, as are exhaust systems made from much less expensive aluminized materials.

Because they're such vital emissions devices, emissions laws limit what can be done with regards to changing catalytic converters. High-performance replacement catalytic converters like the Super High Flow Catalytic Converter from Random Technology, shown, offer a much lower restriction, thus do not restrict power as much as stock converters.

Ever since catalytic converters were first introduced, in the mid-1970s, they have had the reputation of being power-robbing emissions junk. In fairness to the anticlean air crowd, the early pellet-type catalytic converters were highly restrictive and did kill engine power; but by the 1980s, the pellet-type converters had been replaced by the monolith brick-type converters, which we still have today. Brick-type converters present only a relatively modest restriction to airflow and therefore have a minimal impact on an engine's power production. So-called high-flow catalytic converters are available from aftermarket sources, but most use merely the same type of brick just in a larger size and with larger openings and exits.

There are no modifications necessary nor permitted to catalytic converters, nor does the EPA permit replacing them until the converter loses its efficiency, such as through contamination, allowing engine emissions to become dirty again.

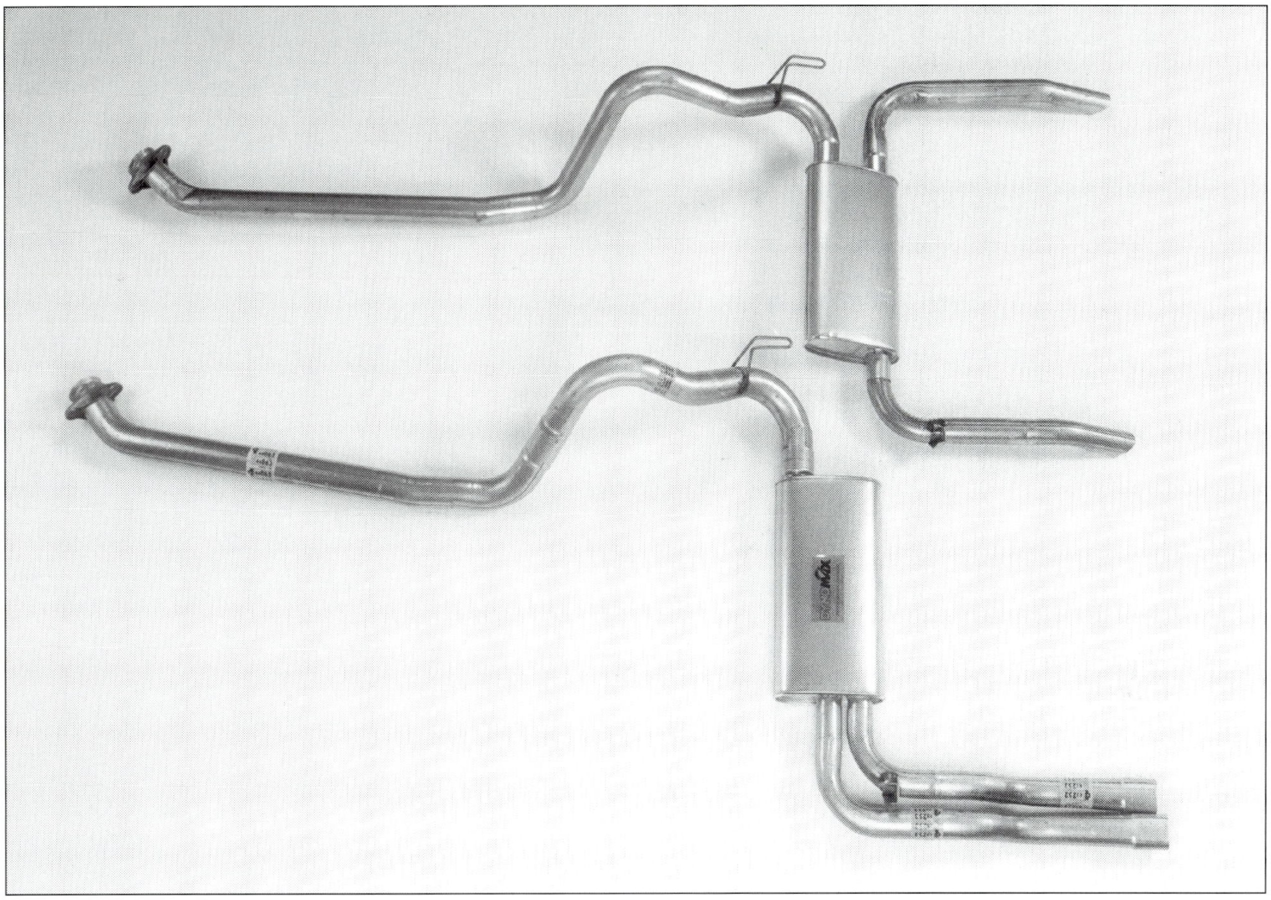

Because of the legal (and physical) complications of changing catalytic converters, the popularity of "cat-back" performance exhaust systems has skyrocketed in recent years. These two late-model Camaro systems are from Walker's DynoMax performance division. The top system is similar to a stock system but features a slightly more efficient muffler. The bottom system, however, has a larger diameter intermediate pipe, increased-flow muffler with a more efficient flow path, and twin tailpipes that exit the same side of the muffler, aiding flow. The performance system is also made of aluminized steel, which resists rust and corrosion much better than nonaluminized exhaust pipes.

While the EPA won't allow you to change your vehicle's exhaust system beyond what it came with in stock form, some Chevrolet models that usually came with a single catalytic converter were available with a high-performance dual-catalyst exhaust system that you could upgrade to. Camaros are the notable example here: A dual-catalyst system was available on TPI engines and eventually became standard on fourth-generation V-8 Camaros. This dual-catalyst system was a terrific power booster, adding as much as 25 horsepower to stock engines. Note that dual catalysts do generally take longer to reach full operating temperature than does a single converter, so if you run dual catalysts and have to subject your vehicle to an emissions sniffer test, make sure you drive the car around long enough to ensure the cats are up to temperature so they can properly do their job.

If you aren't fortunate enough to have the option of upgrading to a dual-catalyst system, you may want to consider upgrading to a high-performance or high-flow catalytic converter to replace your stock converter when it wears out. These high-flow converters really just use a larger monolith brick to increase surface area and thus decrease flow restriction.

Cat-Back Exhaust Pipe/Muffler Systems

Since the Environmental Protection Agency (EPA) makes it illegal to alter any emissions devices on a vehicle exhaust system, manufacturers are focusing on "cat-back" exhaust systems—systems that replace all the exhaust system parts rearward of the catalytic converter, thus keeping the EPA happy, while at the same time providing a noticeable power boost and sound-quality improvement.

Most cat-back exhaust systems employ larger-diameter exhaust pipes, typically with smooth mandrel bends, plus a low-restriction muffler with a larger inlet and one or more large outlets. Additionally, most are available in either aluminized or stainless steel, to ensure the systems last for many, many years.

Exhaustive Recommendations

Ideally, every car would run long-tube, equal-length coated headers, twin cats, and a free-flowing cat-back exhaust system for maximum efficiency. But that's not always possible, nor is it really always necessary.

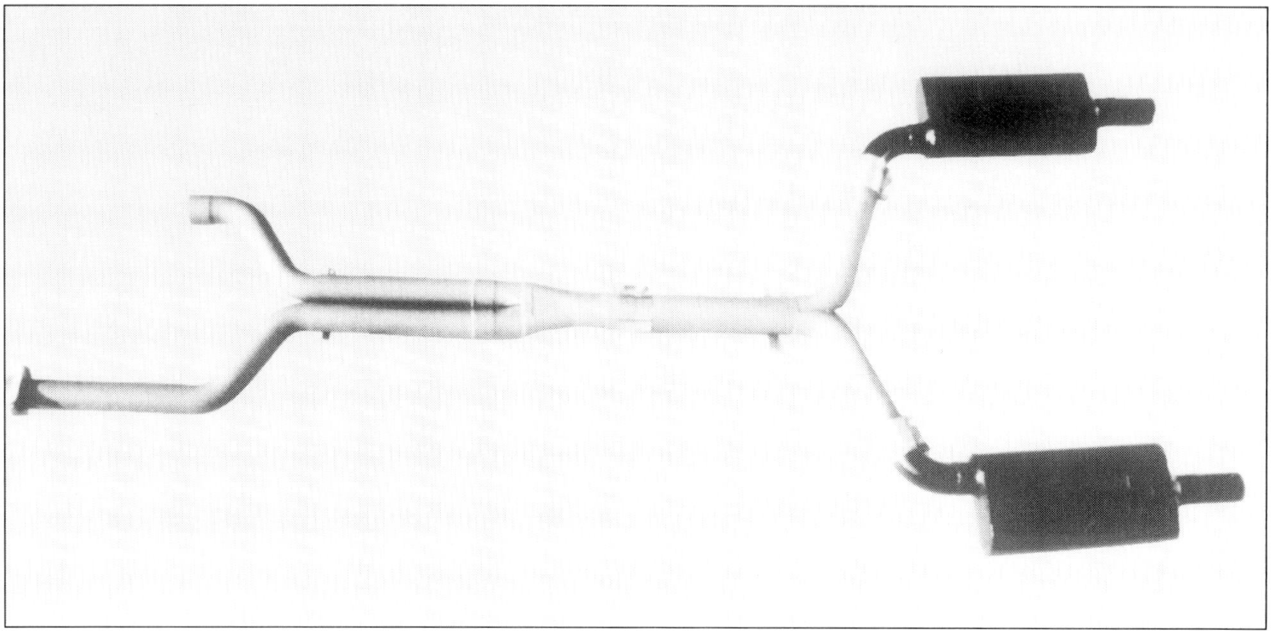

This is a Corvette exhaust system that clearly shows how the intermediate pipes remain separated for much of their length and only join for a relatively short distance before splitting again to feed twin mufflers, which are made by Flowmaster, in this case. *Flowmaster*

Depending on how you use your car and whether you feel it's important enough to spend the time to shoehorn in a set of full-length headers, you will likely find a comfortable system somewhere in between bone stock and the ideal just mentioned in the previous paragraph.

Street High-Performance Driving

Street driving doesn't require an all-out exhaust system, but that doesn't mean there isn't any point in upgrading things a little. Headers are worth the money and time to install, but don't get a set unless they feature a metallic ceramic coating (unless you particularly like the thought of replacing them in a couple of years). If your catalytic converter happens to need to be replaced, either get a high-flow unit to replace it or, if a dual-catalyst system was originally available as an option on the car, go for the dual-catalyst parts. A high-performance cat-back exhaust system won't usually break the bank and will generally provide a modest power increase (but don't expect anything terribly noticeable).

If your car is a daily driver and you highly value reliability, you might just want to stick with the production exhaust manifolds, however, because they rarely ever leak, they're nice and quiet, and they never wear out. Of course, they will cost you a few ponies and they won't exactly help your handling any.

On the off chance that the car you're dropping your TPI motor into can run true duals, don't get too crazy on the diameters of the exhaust pipes. Most 350- to 400-cubic inch street engines will work fine with 2 1/4-inch or 2 1/2-inch exhaust pipes and turbo-style mufflers. Tailpipes don't need to be as large, so 2-inch pipes are usually sufficient. These modest pipe sizes will help keep the flow velocity up throughout the system.

Autocrossing

Power starts to take on a more important role when you're actually entering into some form of competition, like autocrossing, so headers (if allowed by class rules) are much more desirable, if only in shorty form. Equally important to the power boost they'll provide is the weight reduction you'll realize from them.

The rest of the system should be the same as for the street: high-flow or dual catalysts and an efficient cat-back system.

Drag Racing

If you're serious about racing, your only real choice is a set of equal-length, long-tube headers. At the track—and depending on class rules—you can probably uncork the headers by disconnecting the exhaust at the end of the header collector for a few extra horses.

If you are going to all the trouble to run equal-length, long-tube headers, you

True "Duals" Are Most Likely Truly Illegal

Ever since the 1940s (and probably before then), hot-rodders have been fabricating dual-exhaust systems to let their engine's exhale more easily. That tradition has continued through to today, despite the Environmental Protection Agency (EPA) mandates that prohibit altering exhaust systems beyond the configuration provided from the factory. In other words, if a car didn't come with dual exhaust, it can't have dual exhaust.

Fortunately, advances in aftermarket "cat-back" exhaust systems—which replace all components from behind the catalytic converter to the tailpipes—mean there's little need for true dual-exhaust systems. In fact, adding true dual exhaust to a late-model TPI car can actually create more problems than it solves due to loss of ground clearance and other mounting problems.

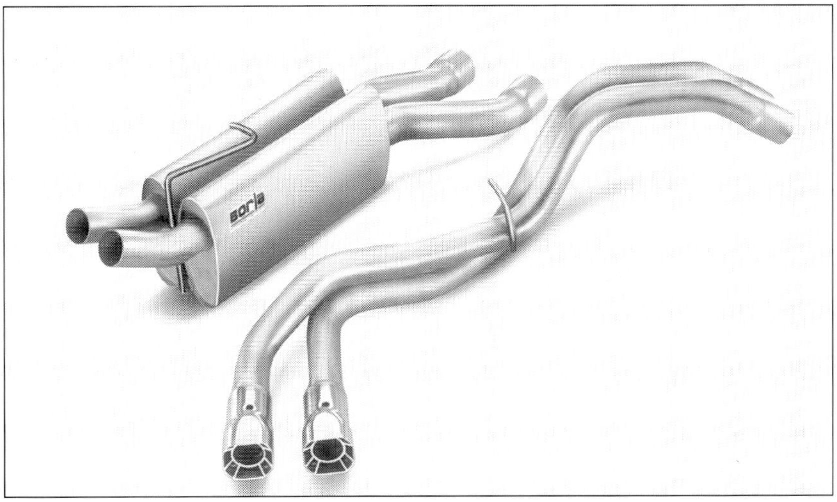

Borla's performance exhaust systems have an excellent reputation for flow, fit, and finish. All Borla systems, like this one for 454-SS pickups, are made of stainless steel for long life and feature only mandrel-bent tubes for minimal flow restriction. *Borla*

also want to consider the tube sizes you'll be running. While street-driven small-blocks respond better to 1 5/8-inch primary pipes, race engines—which typically spend more time at higher rpm—can benefit from 1 3/4-inch or even 1 7/8-inch primary tubes that have the capacity to flow enough air to prevent the engine from choking on its own exhaust. Keeping the primary tubes relatively long—around 30 inches or so—will aid low-end torque too. Collectors have an impact on power too. Good headers will have collectors that measure at least 3 inches in diameter and are about a foot long. Extending them will usually increase low-end torque even further, but you'll have to experiment with the actual length because it varies based on camshaft profile, head flow, and the remainder of your exhaust system (or lack of one, if you're running at the track).

If the class you compete in happens to require you to run through the mufflers, you'll definitely want a cat-back system. A high-flow catalytic converter wouldn't hurt any, either, if yours is in need of replacement.

Road Racing

As with drag racing, power is paramount in road racing, so long-tube, equal-length headers are best. A 1 5/8-inch primary tube that's around 30 inches or longer will help boost low-end torque when rounding tighter corners, as will a 12- to 18-inch long collector. Again, if class rules require a full exhaust system, you'll definitely want to opt for a cat-back system to free up flow. Interestingly, some of the mufflers on the market actually flow *better* than a straight length of pipe. Likewise, catalytic converters only cost a few horsepower, so leaving off either cats or mufflers won't really gain you much, and leaving them on will drive your competition nuts when you run just as well—or better.

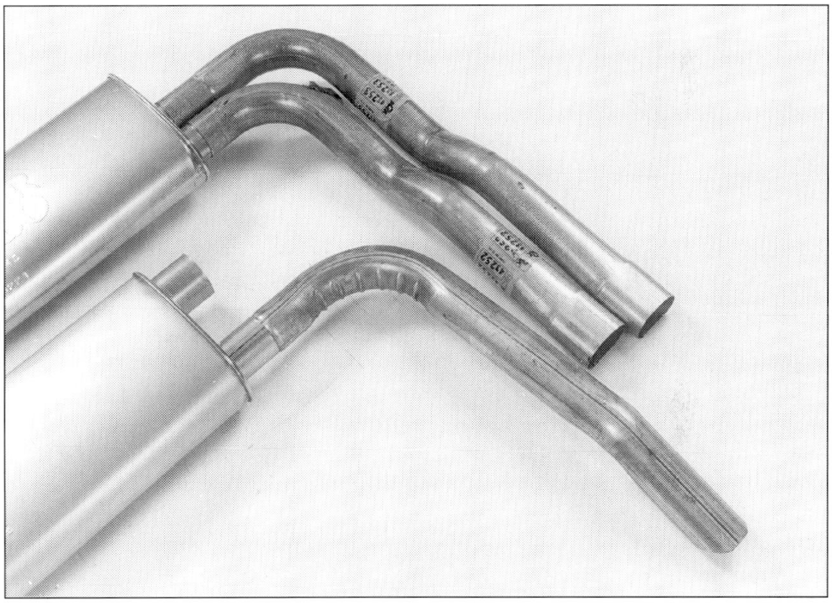

This comparison of two Walker DynoMax systems shows the difference between punch-bent tubing (bottom) and mandrel-bent tubing (top). Not the serrations in the punch-bent tubing, as well as its reduced diameter at the bend. The same area on the mandrel bent tubes is unencumbered with either flow-disrupting serrations or a reduced diameter.

CHAPTER SEVEN

IGNITION SYSTEM

Getting fresh air and fuel into an engine's cylinders won't do any good without an ignition system that's up to the task of burning that air and fuel. See how long your engine runs without spark plug wires and you'll know what we mean (here's a hint: it won't even start without them).

The goal of the ignition system, however, isn't just to ignite the air and fuel mixture, but rather to burn it as completely as possible. The more complete the burn, the more power will be produced.

Over the years there have been a number of significant advances in ignition systems technologies. Until the mid-1970s, ignition points were still common equipment. Then GM's High Energy Ignition (HEI) system did away with points and made high-energy coils common. Then Autotronics Control Corporation revolutionized ignition systems when it introduced its hugely successful Multiple Spark Discharge (MSD) systems that cause the spark plugs to fire multiple times to ensure the air fuel mixture is burned completely.

When General Motors created the LT1 Gen-II small-block V-8 engine for 1992, it replaced the ignition distributor with the Opti-Spark optical crank-triggering system for more accurate ignition timing and, with the LS1 engine, it addressed a long-time problem—insufficient coil-recharge time—by equipping each cylinder of the LS1 with its own individual coil, plus an extremely short ignition wire that results in less energy loss. Older engines that rely on a single ignition coil need that coil to fire 800 times per second for a V-8 running at 6,000 rpm, which means the coil has only 1/800th of a second to step a 12-volt charge in its primary windings up to the more than 50,000 volts the spark plugs are designed to work with. Obviously, splitting up the work among eight coils greatly extends the length of time between firings (to 1/100th of a second), so the coil is much more likely to reach a full charge, and thus the plugs will fire with more intensity, which will contribute to more efficient combustion.

Lighting the fire is your ignition system's sole job. *SplitFire Spark Plugs*

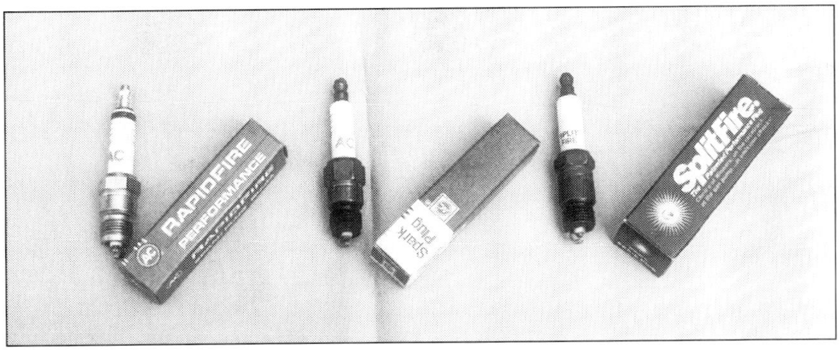

Spark plugs are spark plugs, right? Well, not necessarily. Here are three spark plugs that represent different technology. From left to right: AC's RapidFire, AC's standard resistor plug, and a SplitFire plug.

But, of course, it doesn't take long after GM introduces a new ignition system for someone to figure a way to improve it, so we now have a plethora of performance parts available for your EFI engine's ignition system from aftermarket manufacturers, plus a number of modifications that can be made to tune your ignition system.

Production Ignition System Shortcomings

There are essentially two ways to approach upgrading your ignition system, and it's likely you will employ parts of each. The first steps usually involve improving the performance of the stock system's existing parts, often by replacing them with aftermarket performance parts that offer specific improvements.

The second possibility is to extend the abilities of your ignition system with the addition of various auxiliary performance ignition accessories, such as the MSD multispark boxes mentioned earlier.

The goal, however, is always the same: to deliver a hotter spark that will thoroughly burn the intake charge, thereby maximizing heat production and power.

The LS1 and 1999+ Vortec V-8 systems are so new that few parts are available for them, nor is it clear at this time what (if any) their shortcomings may be. They are state-of-the-art and appear to function exceptionally well.

First- and second-generation small-block ignitions aren't quite so lucky, nor are they easily able to take advantage of LS1-style technology.

The difference between the first-generation TPI engines and the LT1/LT4/L99 engines is that the former use a traditional, rear-mounted distributor to route spark energy to the spark plugs, while the latter employ a front-mounted distributor driven by the camshaft's snout. Note, however, that the Gen-II engines can use a conventional rear-mounted distributor. General Motors even includes an HEI distributor with its carbureted LT4 crate-motors that are sold through GM's Service Parts Operations division through parts dealers.

The biggest problem the TPI first-generation engines typically face is the small-diameter distributor cap used to provide necessary firewall clearance in Camaros and Firebirds. Because of the reduced diameter, it's relatively easy to suffer cross-firing, which is when the spark energy intended for one terminal in the cap actually jumps to another nearby terminal, causing a misfire. The problem worsens as the cap (and rotor) accumulate miles and even trace amounts of contaminants, such as carbon. Because of the wider distances between terminals, large-diameter caps—such as those of the traditional coil-in-cap HEI units—are preferable but won't fit in the third-generation (1982–1992) Camaro and Firebird engine compartments without clearancing the firewall (i.e. taking a large hammer to the firewall).

Gen-II engines equipped with early Opti-Spark cam-driven distributors face a similar problem. The distributors were sealed to prevent corrosion, which seemed like a good idea . . . at first. Unfortunately, the sealed environment tended to ionize the air inside the Opti-Spark unit, resulting in carbon tracking and eventually misfires. Later Opti-Spark systems feature a port that is vented to manifold vacuum to solve that problem.

Additionally, the LT1 and L99 Opti-Spark systems, though newer than the HEI-based systems of the TPI engines, actually feature weaker spark energy. Chevrolet corrected this problem with the LT4 by upgrading the system with a stronger coil; however, the LT4 cannot simply be retrofitted to the LT1 and L99 engines, because the change requires altered PCM program-

Here's a side-by-side comparison of the standard, RapidFire, and SplitFire plugs. Note the differences in the center electrode on the RapidFire plug and the forked tip of the SplitFire.

This is a good-looking standard AC plug, considering it has about 25,000 miles on it. Note the clean, white insulator, which indicates the fuel-injection system is keeping the mixture right. Also note the slight rounding of the center electrode, which indicates normal wear but does reduce spark efficiency. For performance applications, it may be helpful to cut back the plug's tip to the center of the electrode to better expose the spark to the intake charge in the cylinder.

The SplitFire plug's tip is obviously different. The idea behind it is that it better exposes the intake charge to the spark and provides more surface area for a spark to jump to.

ming beyond what the LT1's version of the PCM can handle.

Ignition System Components and Upgrades

Ignition Coil

An ignition coil is a remarkable little device. It takes a lowly 12-volt battery signal and steps it up to as much as 60,000 volts or more! All that voltage is needed, however, in order to jump the spark plug gap strongly enough that the spark can touch off a roaring instantaneous inferno that will burn all of the air and fuel in a cylinder.

Stock coils work well, but when loads are highest and engine rpms are soaring, the coil can't recharge quickly enough to produce an intense spark across the spark plug gap.

Aftermarket coils typically offer increased windings to induce a higher discharge voltage into the secondary windings. Many aftermarket coils also offer quicker refresh times, to improve performance at high rpm.

Distributors, Caps and Rotors

A traditional rear-mounted distributor is used in Gen-I Chevrolet small-block V-8 engines to control where a spark signal is sent (through which spark plug wire) and when.

Production LT1, LT4, and L99 engines do not use a conventional rear-mounted distributor (though they can), but rather the optical triggering device that GM called the Opti-Spark system. The Opti-Spark system is more accurate than a rear-mounted distributor because it is not subject to camshaft vibrations and flex. A crankshaft position sensor tells the ECM/PCM the crank position, and the computer then figures out when to trigger a spark signal, which is distributed through the Opti-Spark system to the ignition wires, and ultimately to the spark plugs. As mentioned earlier, the early Opti-Spark systems were sealed units that developed carbon tracking as a result of the ionization of the air within the unit. Newer Opti-Spark systems feature a port to connect to manifold vacuum to prevent this problem, and older units can be modified to have the same feature.

LS1 engines use an internal crank sensor, as well as a camshaft sensor, to determine crankshaft (and cam) position, then the PCM signals the appropriate coil to fire at the precise moment its spark is needed. While the Opti-Spark system of the Gen-II engines is only able to tell the crankshaft's position, the Gen-III system is much more accurate, being able to tell not only the piston's position in the bore, but the particular stroke, thus it can differentiate between top dead center on the compression stroke and top dead center on the exhaust stroke.

In terms of upgrading or replacing the production distributor, no high-performance replacements are available for the Gen-II Opti-Spark system. First-generation engines, however, can make use of a number of performance distributors that provide more accurate timing thanks to roller bearings for the distributor shaft and a more accurately machined housing that precisely locates the shaft in the exact center of the housing.

Working backward from the spark plugs are the ignition wires, which are the conduits through which spark energy is delivered to the plugs. This is ACCEL's Extreme 9000 wire that is especially effective in high-heat areas.

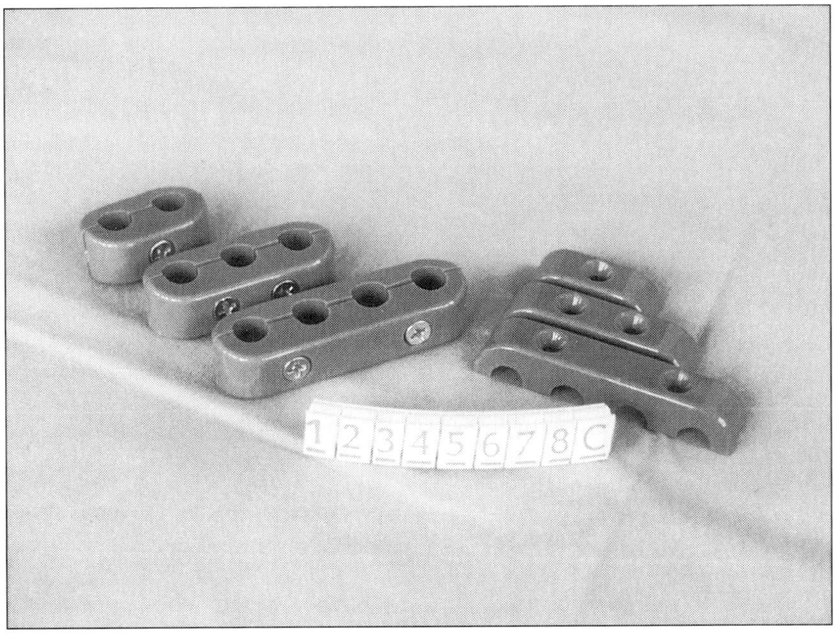

Little things mean a lot. You should always use ignition wire looms to keep plug wires from unnecessarily wrapping over or under each other, which can cause "cross-talk," where spark energy in one wire jumps to another nearby wire, causing a misfire.

Camaro and Firebird owners face the aforementioned space crunch, which dictates the small-diameter cap unless they create enough room for a standard-size HEI or Ford-style distributor cap by pounding back the firewall (note: such a modification is not recommended because it can compromise equipment mounted inside the instrument panel, and even the windshield wiper operation). The stock-sized cap can be retained and still deliver good performance, however, provided that both it and the rotor are replaced frequently, preferably with performance versions that attempt to minimize carbon tracking, plus provide superior metals for the contact terminals.

Note that distributor caps and rotors should always be replaced as units and that parts from one manufacturer should not be mixed and matched with parts from another manufacturer because of differing design and manufacturing methods that can make parts incompatible.

Where's the Computer Info?

Because the engine control management systems are so tightly integrated with the fuel systems on all iterations of Chevy EFI engines, you will find information relating to the ECM, PCM, and VCM, as well as its many sensors, covered thoroughly in chapter 4.

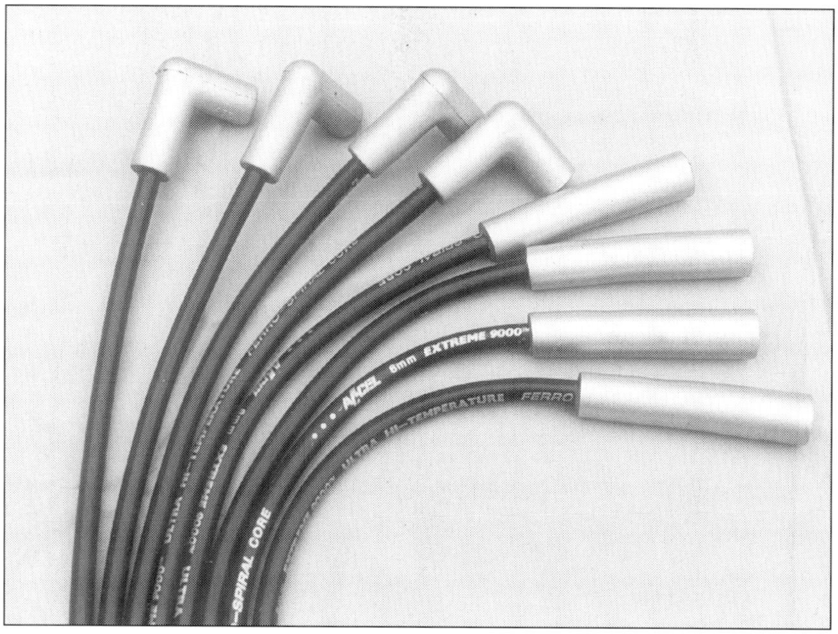

Note the different wire ends of these ACCEL wires. Some engines will dictate use of all angled-boot wires, while others will need all straight plug wire boots. Most stock applications tend to use a combination of the two.

Ignition Wires

An ignition wire is the conduit through which a spark pulse travels to reach a spark plug. Stock wires aren't designed for performance use. They typically feature high resistance—as much as 5,000 ohms per foot—which reduces the current available at the spark plug with which to create a spark.

For performance applications, replacement wires offer lower resistance—some as low as 150 ohms per foot—allowing more spark energy to reach the spark plug. Many also employ superior insulating materials that provide increased protection against spark-induction from one wire to another (known as "cross-firing," which has nothing to do with cross-fire injection fuel-injection systems), plus reduce radio frequency interference, so onboard computers, radios, and communications equipment function properly.

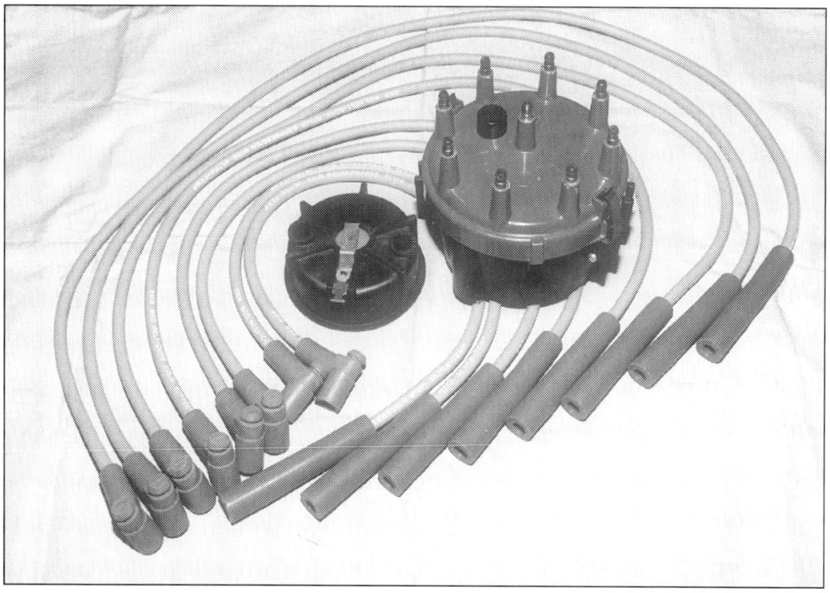

When replacing distributor caps and rotors, always get them from the same manufacturer. Some manufacturers make the tip of the rotor longer and the terminals in the cap smaller, while others do the opposite. Mixing a long-tip rotor with a large-terminal cap could cause the two to contact each other, destroying one or the other or both. Also, make sure wires secure snugly to the cap and plugs. These pieces are from Autotronics/MSD.

Some wires also provide additional protection against high heat, which is especially useful on engines equipped with exhaust headers, which are notorious for melting ignition wires. Ignition wires are available in a number of different sizes, and 8-millimeter wires should be considered a minimum. Likewise, because of all of the interference-sensitive electronics in late-model cars, you should always use RF-suppression wires. They suppress radio frequencies that could disrupt the operation of some vehicle systems, especially radio/stereo equipment.

Spark Plugs

Spark plugs are spark plugs, right? Not exactly. Years ago there were few choices in spark plugs: You replaced old spark plugs with new versions of the same things. Today, however, we have an abundance of spark plug choices. There are choices in heat ranges (which we've had for many, many years), choices in construction materials, electrode designs, tip designs, and more. There are so many choices and so many claims made about why each one is so much better than the others that it can be a frustrating and tiresome exercise to pick the best plugs for your engine.

For years the factory relied on standard AC-Delco resistor plugs, but

ACCEL's 300+ ignition box comes complete with a plug-in wiring harness that makes installation a literal snap (or two). The hardest part is simply finding a spot underhood to mount the box itself.

beginning in 1996 it switched to using platinum-tipped spark plugs to extend general service intervals to 100,000 miles. (Remember when it used to be amazing to have an engine live to 100,000 miles? Now that's the first recommended tune-up interval!) Platinum plugs actually just use platinum electrode tips, which resist wear much better than do standard plug materials. The result is a plug that preserves its gap and spark quality for longer, thus maintaining optimal performance much longer.

A number of companies produce spark plugs with unique design features, such as a forked tip, grooved electrodes, or other innovative elements. While it is possible that these devices will add power to your EFI engine, it's also likely that a fresh set of standard plugs will provide much of the same benefits. In other words, the high-tech features really don't provide much more performance than a set of new, standard plugs.

How you use your engine and the performance equipment that it has can affect the specific spark plug specifications necessary to maintain optimal efficiency and power. One typical adjustment involves the plug heat range: how quickly or slowly the plug can cool itself. General Motors typically specifies a rather hot plug for production use, given that most vehicles are going to be driven on lots of short, around-town trips that may not always be sufficient to burn the spark plugs clean of soot and carbon. But in racing applications—particularly road racing, where runs are long and cylinder temperatures remain high for extended periods of time—the stock plugs may not cool quickly enough and may actually begin to melt. So, it may be advantageous to switch to a cooler plug heat range than stock.

There would rarely be a reason to switch to a hotter-than-stock plug, especially if you upgrade other aspects of the ignition system.

Auxiliary Ignition System Accessories

High-performance and racing activities often place special demands on an ignition system and may benefit from the use of certain aftermarket ignition products designed to increase spark output and thus maximize engine power and efficiency. The

Combustion Basics

Many people understandably misunderstand the combustion process in an internal combustion engine. It's a complex process that is dependent upon a number of factors, including the fuel you burn, the temperature (and thus density) and humidity of the air, piston top and combustion chamber shape, spark plug design and location, cylinder head material, engine load and rpm, and other factors.

One of the most important concepts to grasp is that power isn't generated by an "explosion" of the fuel and air when ignited by the spark plug. Power is generated by the heat produced by the combustion process. In fact, the air and fuel technically don't explode—they take time to burn, not a long time, but time nonetheless.

As the air and fuel mixture burns, the gases in the combustion chamber become extremely hot and expand, which forces downward against the piston, which, in turn, causes the crankshaft to rotate. To get the most power, however, the combustion process needs to start early enough so that the piston isn't too far down the bore (and descending farther) when the combustion process finishes. Otherwise you will be wasting power, because the volume in the cylinder at that point will be great enough that expansion of the gases will have little or no effect on the piston's descent. On the other hand, if combustion starts too early before top dead center, the piston's ascent will be slowed by the increased pressure that results from the heated gases in the cylinder, and you'll lose power. So it's a fine line between starting the combustion process too early or too late, and that "best compromise" point varies from engine to engine based on factors such as those mentioned above.

The best-case scenario is to delay ignition until as close to top dead center on the compression stroke as possible, then burn the mixture as quickly and thoroughly as possible. This method maximizes compression of the air and fuel mixture plus minimizes the counterproductive forces generated by combustion pre-TDC, yet yields maximum power by completing combustion quickly.

To do this, however, your ignition system has to be in optimal condition, because the compressed air and fuel mixture is harder to ignite, so the initial spark must be intense. In addition, to ensure the air/fuel mixture is thoroughly and quickly burned, the duration of the spark must be increased, or multiple sparks must be triggered in rapid succession. In practical application, multiple sparks work better at low- and mid-range rpm, but at higher rpm there isn't enough actual time to trigger multiple sparks. One extended-duration spark provides the same effect: a more efficient, shorter combustion cycle.

most common of these are multiple-discharge spark systems and rev-limiters, but there are others depending upon your application.

Multiple-Discharge Spark Systems

Aftermarket multiple discharge spark boxes, as their description implies, deliver a series of spark pulses in rapid succession to more completely burn the air and fuel in a cylinder. While the initial pulse ignites the air and fuel and is therefore responsible for burning most of the air and fuel, follow-up sparks can ignite pockets of air and fuel that may have escaped the combustion process, resulting in greater overall efficiency. In addition, many of these multiple-discharge spark boxes deliver a single, extended-duration spark discharge at higher rpm, rather than multiple discharges, because high engine speeds do not leave sufficient time to fire a plug multiple times, but a longer-duration single spark can provide the same benefit.

Installation of a multiple-discharge spark box is a recommended

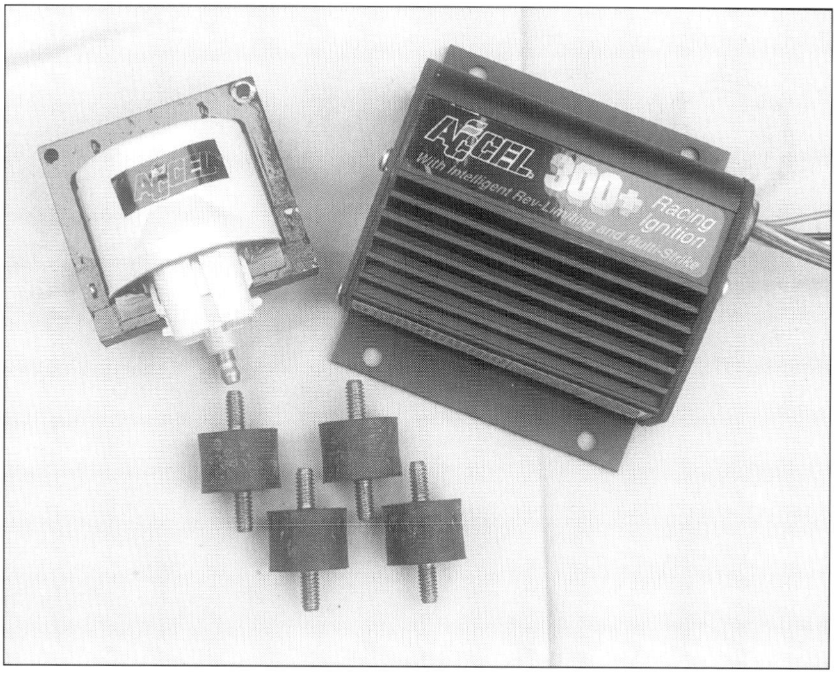

Upping your battery's 12.6 volts to around 50,000 to 60,000 is the job of the coil, like this ACCEL Super Coil (left). Ignition amplifiers like ACCEL's 300+ Racing Ignition box (right) provide multiple spark strikes at low- to mid-rpm ranges, and at high rpm provide a single, extended spark in order to create a more complete burn of the intake charge. The four black items at the bottom are vibration-suppressing rubber mounts for the 300+ ignition box.

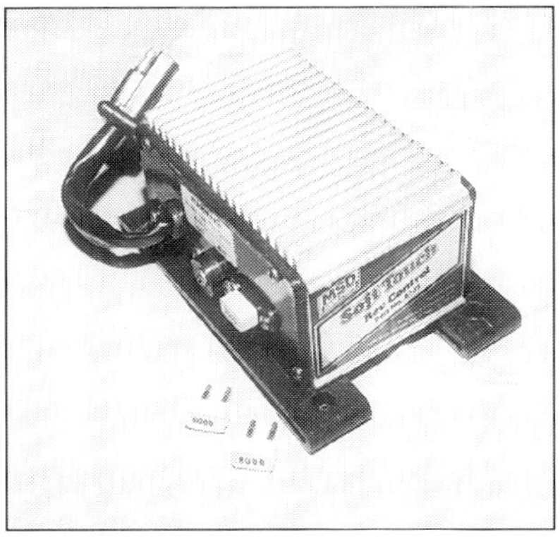

Another worthwhile ignition component is a rev-limiter, such as MSD's Soft Touch Rev Control. These devices prevent you from over-revving your engine (on acceleration) by cutting out ignition signals at a user-defined rpm. MSD's unit uses small plug in rpm modules for quick, easy changes.

upgrade for all forms of driving—street, drag racing, autocrossing, and road racing.

Rev-Limiters

Whether racing or just playing around on the streets, it can be all too easy to get wrapped up in the moment and forget to keep an eye on the tachometer to avoid exceeding the engine's redline (maximum recommended rpm). Over-revving the engine can damage a number of internal engine components, most notably pushrods and valve springs, though damage deeper in the motor is possible too. A rev-limiter can eliminate excessive engine rpm as a concern by preventing the engine from exceeding a specific rpm, which is typically user-programmable. When the engine reaches the specified maximum rpm, the rev-limiter will begin dropping spark pulses, to keep the engine from speeding up (or even sustaining the speed it has).

Sophisticated rev-limiters offer different rpm settings for different situations. In drag racing, for example, it's often useful to perform a burnout to heat and clean the tires. Then, when staged, awaiting the green light, a different maximum rpm is needed. Finally, during the run, a still different rpm is needed to maximize power without sacrificing engine components or reliability. While some systems only provide a single maximum rpm, others allow you to accommodate multiple settings.

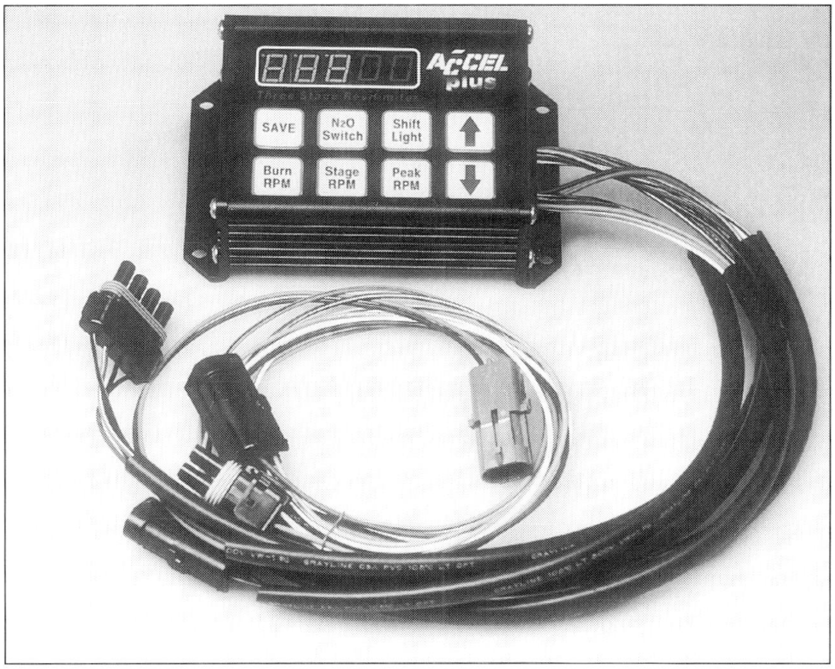

ACCEL offers a three-stage rev limiter that allows you to set different maximum rpm for burnouts, staging, and peak rpm to cover three different needs of competition driving, such as drag racing. It comes with an easy-to-connect wiring harness for plug-and-play installation. *ACCEL*

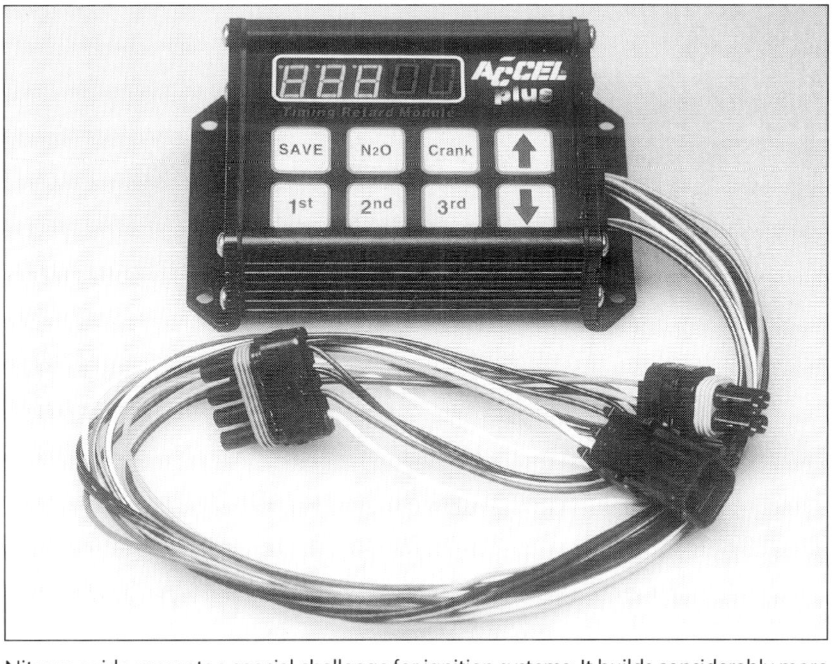

Nitrous oxide presents a special challenge for ignition systems: It builds considerably more cylinder pressure, which requires different ignition timing, but it only needs it when the nitrous is activated. ACCEL's timing retard module allows you to have your ignition automatically increase spark lead whenever the nitrous is triggered. *ACCEL*

Remember, however, that rev-limiters won't prevent accidental over-revving that may happen due to downshifting at high speed. For example, if you attempt a full-throttle up-shift to third from second but accidentally hit first gear, engine rpm will quickly race beyond your engine's redline, potentially

A timing computer, like ACCEL's ignition timing computer, allows you to program in a specific ignition curve for a distributed engine, including automatic changes when boost pressure reaches a user-determined point or when nitrous is activated. *ACCEL*

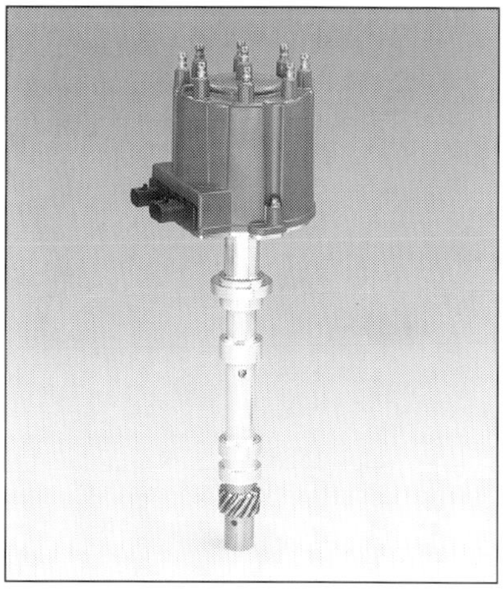

Factory distributors work well, but they're not exactly precision-made parts. The distributor shaft can wobble considerably within the housing, leading to erratic spark distribution. ACCEL's Billet F-Body distributor features the reduced cap size necessary to clear Camaro and Firebird firewalls, plus it features roller bearings that keep the shaft perfectly centered at speeds as high as 10,000 rpm. It's a direct drop-in replacement for a stock F-body distributor. *ACCEL*

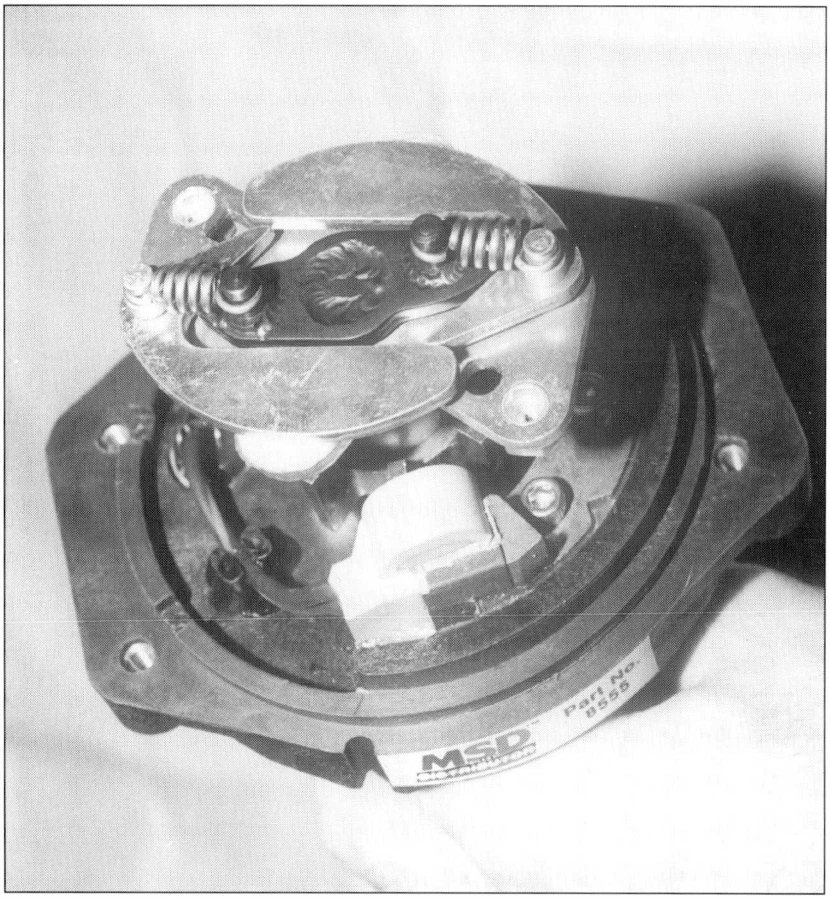

This is a close-up of the mechanical advance mechanism of the MSD distributor. As the shaft spins, centrifugal force pulls weights away from the center of the shaft, which shifts the location of the pickup module to advance the timing. The springs provide some resistance to keep the weights near the shaft at low rpm, for less total advance.

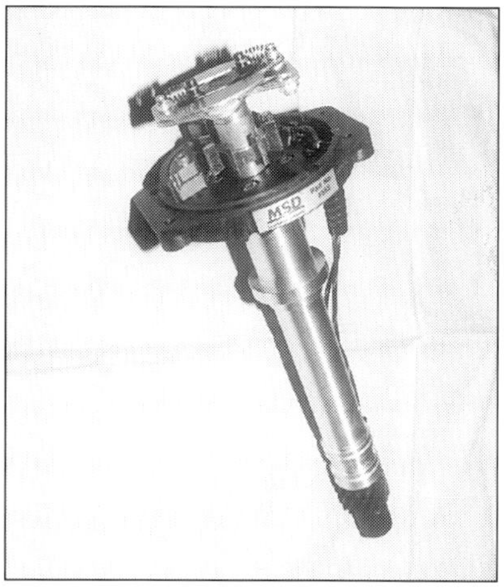

If firewall clearance isn't a problem, a billet distributor like this MSD unit is an excellent solution. It uses larger diameter caps and rotors, which reduce the likelihood of spark scatter and misfires.

causing extensive (and expensive) internal damage, because neither an ignition nor fuel cut-off-based system has any effect in those cases.

Timing Retard Systems

Timing retard systems are used to retard the timing at certain times, such as when running nitrous oxide, or simply when running extended runs that cause the engine to run hot. Some systems use a computer to delay timing events, and most of these types of units allow you to specify the degree of retard to dial-in. Other systems use a fixed secondary triggering mechanism in the distributor. When running off the primary triggering system, the timing is as you set it, but when you switch over to a secondary ignition system, the timing is altered by a few degrees.

Back-Up Ignition Systems

NASCAR fans are probably familiar with drivers suddenly slowing on the race track, only to pick up speed again as though nothing had happened. When asked about it, the driver or crew chief often mutter something about having to switch ignition systems. What they're referring to is the practice of equipping a car with a back-up ignition system that features duplicates of the failure-prone components of the ignition system—namely the coil, multiple-discharge spark box, rev-limiter, and other components—just about everything except the distributor and spark plug wires. A toggle switch is used to select which ignition system is used. Some systems are available with the secondary ignition system triggering mechanism advanced or retarded a few degrees to allow the system to not only serve as a back-up system, but an on-track performance tuning device as well.

Mounted to the front of the water pump of this LT1 is the Gen-II engine's distributor, known as the Opti-Spark system. It's driven off the snout of the camshaft and generally works well, though early systems were not vented to engine vacuum and thus resulted in ionized air and carbon tracking.

A small but important part of a distributor is its gear, which mates with a gear at the rear of the camshaft. Flat-tappet cams are typically hardened cast iron and need a cast-iron distributor gear. Roller cams are steel, though, and a cast-iron gear would be chewed up and would ruin the cam too. So a soft, bronze gear is needed to prevent damage to the cam. The bronze gear will last a good long time, unless subjected to addition stress, such as with a high-volume or high-pressure oil pump. *ACCEL*

All your ignition hardware is reliant upon you setting the timing accurately, but factory timing indicators can be off by as much as a few degrees. Adjustable timing indicators, like this ACCEL unit, allow you to accurately establish top dead center between the compression and power strokes during build-up, so you can accurately adjust the timing later. *ACCEL*

CHAPTER EIGHT

COOLING SYSTEM

Your TPI engine's cooling system is designed to keep a stock engine cool under a wide variety of circumstances, including enthusiastic driving. But if you've modified your engine to create more power, it will tax your cooling system more heavily, and if your cooling system isn't up to snuff, your engine could overheat, potentially causing serious internal damage.

The first thing to understand about TPI engine cooling systems is that there is a world of difference between the traditional first-generation cooling system and the Gen-II cooling system. Namely, the Gen-II engines use what is called a "reverse-flow" cooling system, which sends fresh coolant coming from the radiator to the cylinder heads first, then down into the block—opposite of how a first-generation engine directs its coolant. There are advantages and disadvantages to each, and we'll cover those in more detail later in this chapter. The third-generation small-block—including the LS1 and 1999 Vortec truck engines—cooling system is somewhat of a hybrid of the first- and second-generation systems: They're standard flow, but with several notable improvements that stemmed from research and experience earned with the Gen-II motors.

While most people logically think of an engine's *primary* cooling system components—coolant/antifreeze, water pumps, radiators, fans, and thermostats—when discussing cooling systems, there are other components and systems that affect your engine's ability to keep its cool, such as its oiling system, its ignition system, and even choices of specific components like cylinder heads. These are considered to be *secondary* cooling system components.

Primary Cooling System Basics

First-Generation Small-Block Cooling Systems (TPI, TBI, Pre-1999 Vortec)

Between 1955 and the Gen-II engine's debut in 1992, there had been virtually no significant changes to the cooling system of small-block Chevys beyond the 1969 introduction of the long-neck water pump (which was really more for better accessory drive systems than for improved cooling). In fact, the

When the subject of cooling system upgrades comes up, it's only natural to think of replacement radiators, like this aluminum unit from Griffin Racing Radiators. Extensive use of aluminum not only makes the performance radiator lighter than a production radiator, it's also better at dissipating heat and will usually last much longer too.

Many production radiators have plastic end tanks, like this one. Unfortunately, the plastic tanks often fail and develop leaks, especially where they mate with the radiator core. Production radiators typically do not feature enough rows of tubes to provide sufficient cooling when continuously subjected to high heat, such as during racing.

most noticeable change was the switch in the 1980s to electric fans instead of engine-driven fans in performance cars, like the Corvette and Camaro.

The system is simple and effective, as attested to by its more than 40 years of trusty service. A water pump is turned by a belt driven by the crankshaft. The pump draws coolant through a hose from the radiator and forces it into the block through four openings in the face of the block, two to each side of the timing chain cover for each bank of cylinders. The coolant circulates around the cylinders (except on 400-inch small-blocks, which have siamesed cylinders) then travels into the cylinder heads. From the heads, the coolant passes into the intake manifold at the front of the engine, through a thermostat (once open), and back through a hose to the radiator.

To cool the coolant, a water pump–mounted, engine-driven multi-blade fan, or one or more electric fans, draw air through the radiator. While the engine-driven fans have a thermostatic drive hub that engages when the temperature of air flowing through the radiator dictates, the electric fans are controlled by the ECM based on coolant temperatures; they're on when they need to be (and the ignition switch is "on") and not when they're not needed.

The Gen-I systems were under modest pressure except for the overflow or surge tank, which was simply at atmospheric pressure.

Forty-odd years of service prove beyond a doubt that the system really does work well, but it isn't without its flaws. The most significant is that the coolant heats up considerably in the block, so by the time it reaches the heads it can't absorb much more heat, so the combustion chambers stay pretty hot, which actually promotes detonation, thus compression ratios need to be kept relatively modest. Another flaw of the system is that it has an abundance of mating surfaces, each of which poses a threat of developing a leak. For example, here are a few of the potential leak spots in the Gen-I engine: two gasketed surfaces for the water pump, water pump radiator hose neck, water pump heater hose fitting, water pump rear cover, cylinder head gaskets, intake manifold gaskets, heater return line fitting, thermostat gasket, thermostat housing radiator hose neck, radiator inlet/outlet necks, overflow hose, radiator cap, core/"freeze" plugs, and even some of the cylinder head bolts that pass through the block's deck surface into the coolant passages. That's a lot of potential leaks.

The biggest improvements to the first-generation small-blocks are made with better water pumps and better radiators. It's worth mentioning that switching to aluminum cylinder heads will also greatly affect cooling, since aluminum gives up heat much more quickly than cast iron, making aluminum heads less prone to detonation with higher compression or more total ignition advance. There is also a complicated modification

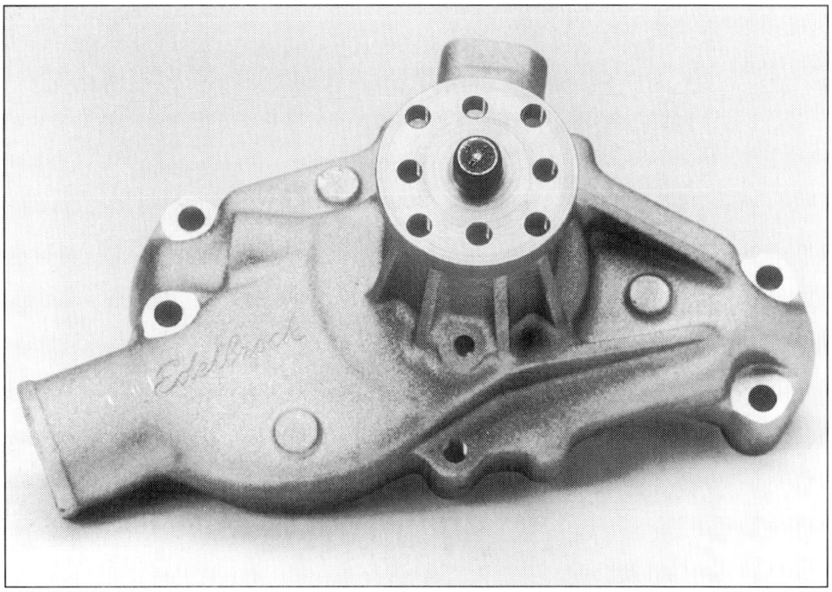

The water pump is another weak point on many production engines, particularly first-generation small-blocks and all big-blocks. Edelbrock's Victor series of aluminum water pumps were designed to withstand the severe demands of NASCAR racing. To minimize the chance of failures, the unit features oversized needle bearings, a cast-iron directional impeller for improved flow rate and direction, and the heavy-duty but lightweight aluminum housing that also helps dissipate heat. They're about triple the price of a standard replacement water pump, but they're worth every penny, at least in terms of peace of mind. *Edelbrock*

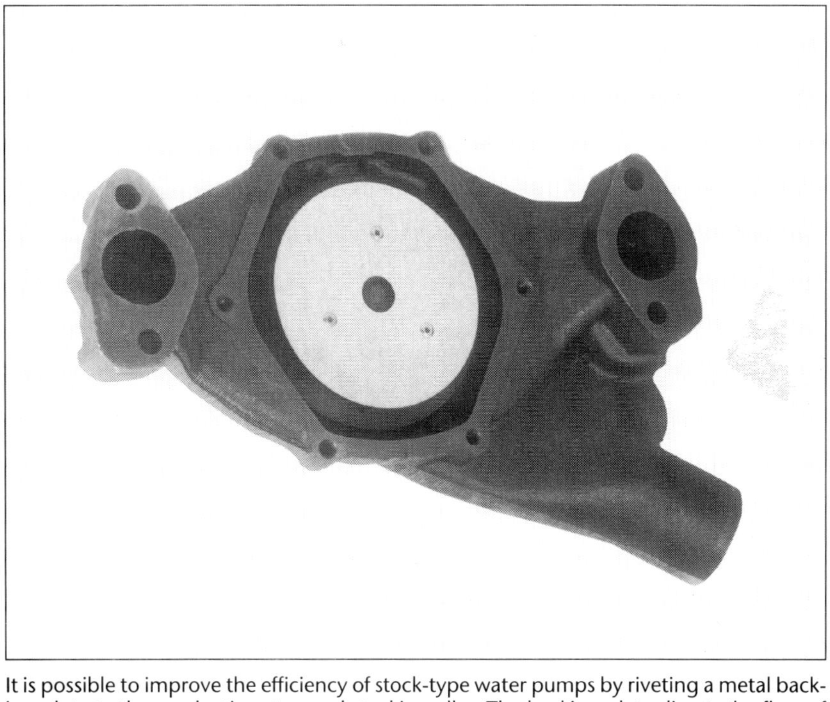

It is possible to improve the efficiency of stock-type water pumps by riveting a metal backing plate to the production stamped-steel impeller. The backing plate directs the flow of the coolant better, for improved pressure and flow through the engine. If you're not up to fabricating your own disc and attaching it to your stock pump, you can simply buy one with the modification already completed from Milodon. *Milodon*

that can be made to cylinder heads (aluminum and cast iron) on engines subjected to endurance-style competition. The modification involves tapping into the coolant passage between the middle two exhaust ports, because the adjacent exhaust ports retain far more heat than do the end exhaust ports. Then a coolant line is run from this new outlet to the coolant crossover passage at the front of the engine, where it is eventually sent through the thermostat to the radiator to be cooled. This modification is rarely useful in anything but professional-level competition engines, however.

There are two basic housing designs with the water pumps: the pre-1969 short pump and the 1969-and-up long pump. Each of the housings is available in either cast iron or aluminum. The choice will depend at least in part on how concerned you are about your engine's weight. In addition, pumps intended to be run by an old V-belt spin clockwise, while pumps for serpentine belt systems turn counterclockwise. In addition, a few aftermarket companies manufacture racing-inspired water pumps (usually in aluminum only) with other upgrades, including larger bearings (to increase durability), O-ring rear cover seals (for better sealing), and different impeller designs for increased flow volume and velocity.

Standard water pump impellers are typically thin stamped steel and basically resemble a fan blade; however, some high-performance aftermarket pumps feature a stamped impeller with a flow-directing backing plate, or even a cast-iron impeller with more efficient vanes. Such pumps are pricey but worth the added peace of mind.

The other change that TPI engine builders need to devote some thought to is which thermostat to run. First of all, you may have heard that you should never consider running without a thermostat because coolant may flow through the engine too quickly, exiting before it has had the ability to absorb much heat. It's true that you shouldn't run without a thermostat, but it's actually for the opposite reason: Without a thermostat, more coolant flows through your engine in a given period of time than with one, and since a higher volume of coolant logically increases the amount of heat that can be absorbed from the engine, it's easy to understand how running without a thermostat can actually make your engine run too cool. Similarly, you can't simply run 160-degree (Fahrenheit) thermostat as in old muscle cars, because the ECM/PCM will interpret a 160-degree water temperature as a sign that the engine isn't fully warmed up, so it will prevent the system from going into the more efficient closed-loop operating mode. The fix for this is a thermostat that runs slightly hotter—170 to 180 degrees Fahrenheit.

The specifics about the radiator you run will be largely determined by the vehicle in which the radiator will be installed. There's more information on radiator selection criteria later in the chapter.

Gen-II Small-Block Cooling Systems (LT1, LT4, L99)

As we learned earlier in this chapter, part of what made the Gen-II LT1 seem so radically different when it was released in 1992 Corvettes was the engine's reverse-flow cooling system. By routing cool, fresh coolant to the cylinder heads first, the operating temperature of the heads—and thus of the combustion chambers—was lowered, so the likelihood of hot spots forming was reduced. That helped minimize detonation. This allowed engineers to raise the static compression of these engines for more power. The LT1, for example, featured 10.25:1 compression, and the LT4 actually ran fine on 92-octane premium unleaded pump gas with its 10.4:1 squeeze.

To make reverse-flow cooling possible, Chevy had to redesign the water pump, block face, and cylinder heads. The water pump sent the coolant through larger, revised ports into the block then directly up to the heads. The coolant circulated around the combustion chambers then down into the block, then back out through the larger ports in the face of the block and into the water pump. Note the coolant never enters the intake manifold. This was done for two reasons: first, by keeping potentially hot coolant out of the intake manifold, it minimized the tendency for the intake air charge to heat up, and thus become less dense; second, by not sending the coolant into the intake manifold, the chance of a coolant leak at the manifold was eliminated.

The remainder of the Gen-II primary cooling system was pretty much standard-type parts: two hoses and a radiator, with a feeder hose to the overflow tank and electric fans.

The Gen-II water pumps are not interchangeable with first-generation small-block water pumps. They are, however, far better units than the production first-generation water pumps. With aluminum housings, a high-volume directional impeller and beefed-up bearings, the LT1/LT4 pumps are generally well suited to even high-performance use.

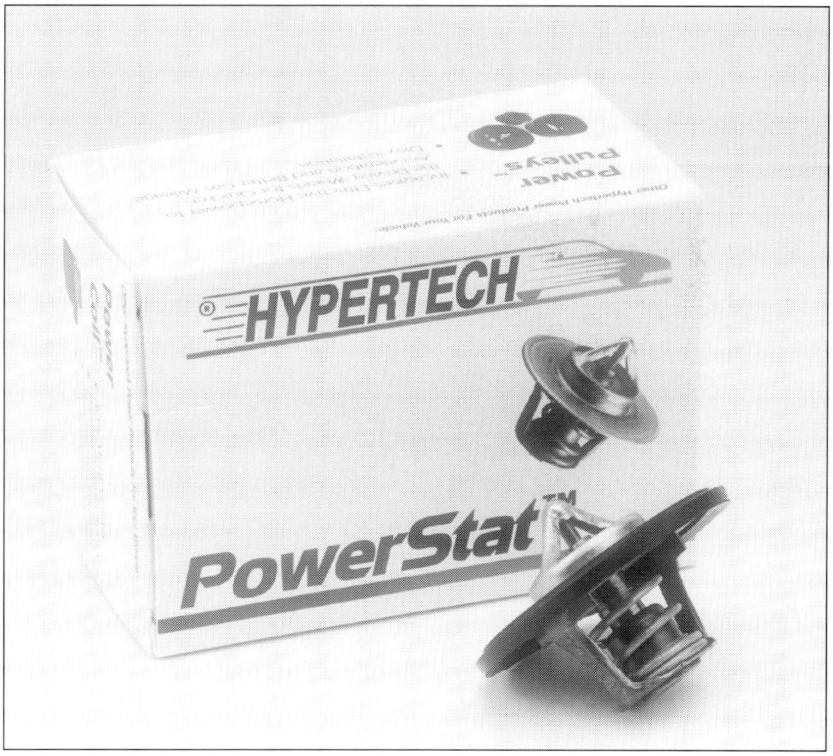

EFI engines are particularly finicky about thermostats, because they depend on coolant temperature sensors to determine when the engine is thoroughly warmed up. While 160-degree thermostats were the hot—or rather, cool—ticket with carbureted engines, that wouldn't allow an EFI system's ECM to go into the closed-loop mode, so performance and emissions would suffer. HyperTech and other companies offer a 170-degree thermostat that results in an only marginally higher coolant temperature but allows the ECM to go into closed-loop mode for peak performance. *HyperTech*

A thermostat's opening temperature is usually engraved on the top of the unit, as shown here. The "170" indicates a 170-degree opening temp.

Some thermostats, however, have the opening temp engraved in the bottom of the unit like this 170-degree piece.

Either fortunately or unfortunately, depending upon your perspective, there really aren't many upgrades you can make to the Gen-II small-block cooling system. Since the water pump housing is already made of aluminum, it is both lightweight and able to dissipate heat well, plus it produces excellent flow through the engine, so there is no need to replace it. The electric fans are generally more than able to provide the necessary cooling. In fact, the only area that could benefit from improvements is the radiator—especially on vehicles equipped with radiators with plastic side tanks, which are notorious for failing at the most inopportune times.

Third-Generation Small-Block Cooling Systems (LS1, 1999+ Vortec V-8s)

When it came time to design the Gen-III small-block's cooling system, Chevy had lots of real-world experience to draw from, with both standard-direction flow and reverse-flow, as well as with a number of different types of traditional modifications and parts. Engineers knew what worked and what didn't, so they set out to combine the best all-around features to create the ultimate small-block cooling system.

They returned to the traditional small-block Chevy flow direction—through the block first, then on to the heads—but modified the design to avoid flowing through the intake manifold, as they had done with the Gen-II system. They also enlarged the ports in the block to match the coolant flow capacity of the new double-volume water pump.

The cylinder heads on all Gen-III small-blocks feature symmetrical ports, which prevents two exhaust valves from being side-by-side, as in Gen-I and -II heads. In addition, all Gen-IIIs (except the Vortec 6000) feature aluminum cylinder heads, which further reduce the heat in the motor by virtue of the alloy's superior heat dissipation properties. These two changes made reverse-flow cooling unnecessary.

The other big change for the Gen-III engines is the move to a fully pressurized system, including the overflow tank. Fully pressurized systems are so-called zero-loss systems, meaning that they aren't supposed to lose any coolant—ever. If they do, there's a problem, such as a failing head gasket, a leaking radiator, or some other malady. The Gen-III engine management systems use this to their advantage to monitor not only the coolant temperature but the volume of coolant too. The engine management systems can alert the driver when it detects a possible problem, giving the driver the ability to have repairs made before engine damage occurs. In the 1999 Vortec V-8-equipped trucks, the engine management controls can even allow the engine to continue running without any coolant, to allow drivers to make it to a service center for repairs. (For more on this interesting feature, see the sidebar "Air-cooling a 1999 Vortec V-8.")

Because the Gen-III system is so well engineered, there are, again, few improvements that can be made, beyond higher performance, heavier duty radiators and possibly auxiliary fans.

Primary Cooling System Components

Radiator

When people think of the cooling system, one of the first two things that usually come to mind is the radiator (the other is the coolant). Though technically not a part of your engine, your engine wouldn't live very long without the radiator, because the radiator is what takes the hot coolant from the engine and cools it off, by sending it through thin tubes (arranged in rows) over which air flows. Additionally, thousands of tiny fins zigzag back and forth between the rows, absorbing additional heat from the tubes and shedding that heat into the air flowing past them.

Generally speaking, the more rows and columns of tubes a radiator has, the better it can dissipate heat. But there is a practical limit to how thick a radiator can be before it reaches a point of diminishing returns. For instance, a four-row radiator is generally considered to be the thickest radiator useful in a street car. Any more and the radiator becomes so thick that it actually acts as a wall, impeding air from flowing through it. Even if air could flow through it easily, the air would be so hot from the heat given off by the first four rows of tubes that the fifth row couldn't dissipate much heat.

The radiators installed at the factory in TPI-equipped cars, like third- and fourth-generation Camaros, used plastic end tanks that act as large reservoirs to hold coolant before and after it flows through the cooling tubes. Unfortunately, the plastic tanks are notoriously problematic, frequently suffering leaks, cracks, and other troubles. The finned radiator core is made of copper and has

Air-Cooling a 1999 Vortec V-8

Though it's really more of a miraculous feat of computer engineering, the fact that the 1999 Vortec V-8s can operate without any coolant bears at least a passing mention in any chapter on small-block cooling systems—if only because it's so darned neat.

Thanks to the electronic sensors and the powertrain control module, it's possible to have the engine protect itself when things go awry. Of course, the easy thing to do would have been to simply shut down the engine when it detects severely low engine coolant, but since engineers couldn't predict whether doing so would be safe (picture someone on the highway and their truck all of a sudden dies on them; that's a good way to cause an accident!) the GM powertrain engineers devised a method of allowing the engine to continue to run without coolant, to allow a driver to get to a service center to correct the problem, rather than become stranded miles from nowhere.

The particulars are a bit complex, but the basic gist of things is that when coolant level drops to a specified level, the PCM will kill off one bank of cylinders and continue to run on only four cylinders. The nonoperating cylinders will, of course, continue to pump air in and out of themselves, and that air (and the fact that engine power is severely decreased) keeps the engine cool.

This sort of feature would have been impossible in the old days, but thanks to EFI and computerized engine management systems, the unneeded fuel injectors simply don't spray fuel, and the ignition coils for those cylinders don't fire.

As cool as this feature is, it's not exactly something you can adapt to your vehicle, unless, of course, you swap not only a 1999+ Vortec V-8 into it, but also all the computer controls and sensors.

And even if you do own a 1999 Vortec V-8, you're probably better off not testing this way-cool feature.

It's easy to overlook your radiator hoses and fan belts, but they need to be replaced often—at least every two years or 24,000 miles for street drivers, but more often for competition vehicles. In fact, for competition vehicles it may be wise to replace portions of the radiator hoses with steel tubing, which is less likely to rupture.

Keep It Cool. Keep It Running.

One of the worst things you can do to your engine is shut it off. When you kill the engine, the water pump stops turning and that means any coolant in the block sits there, boiling as it tries to absorb the heat from the engine, which now has no good way to dissipate it.

Without the coolant flowing through the system, the engine will develop hot spots in some areas where heat dissipates more slowly, such as around the middle two cylinders of each bank, while the outer two cylinders on each bank may cool relatively rapidly. These changes in cooling rates can fatigue the engine's block or cylinder heads, potentially causing them to warp or even crack.

That's the bad news. The really bad news is that there are really only two ways to avoid that: First, you could simply not start the engine in the first place, but what fun is that? Second, you can make sure that just because the engine gets shut off, the coolant flow doesn't. Sadly, the only way to accomplish that is to use an electric motor to drive the water pump instead of the crankshaft. Currently, however, the only electric water pump drive systems available are for drag racing use only, because the systems do not vary flow rates based on need, but rather stick to a predetermined speed, which may not be fast enough for your driving conditions. Likewise, the electric drive systems are not designed to be used for long periods of time.

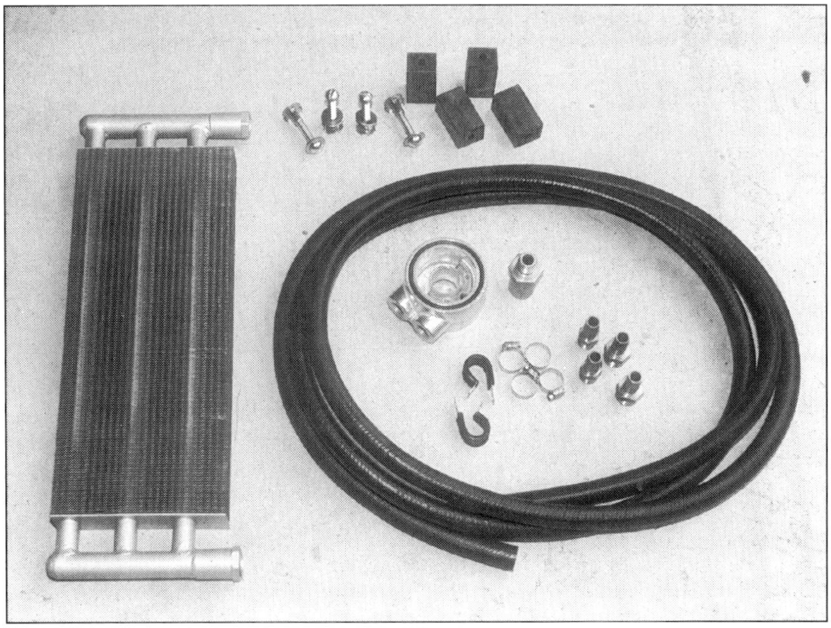

Cooling systems for performance engines involve much more than just the radiator, water pump, and coolant. Oil coolers (for both the engine and transmission) are wise investments, because cooler oil has a higher viscosity and thus can protect your engine better. This kit, from Hayden, features a filter-mount adapter, the necessary fittings and hoses, plus the generously sized radiator that actually provides the cooling power.

no particular drawbacks, but it doesn't have any particular strengths either. The number of rows used varies depending on the engine and options ordered.

Most aftermarket performance or racing radiators are made of aluminum, which weighs less than copper radiators and dissipates heat more quickly, making it more efficient. Along with the aluminum core, most aftermarket radiators feature aluminum end tanks, which virtually eliminate leaks, compared with the plastic tanks used on many production radiators. The aftermarket units are generally available in various thicknesses (rows of cooling tubes), with various cooling tube sizes, and even different fins-per-inch counts, allowing you to customize your radiator to suit your exact requirements.

Water Pump

The water pump forces coolant through the engine by means of an impeller driven by the crankshaft. First-generation V-8 engines turn the water pump with either a V- or flat serpentine belt, depending on the particular year. Anyone who's ever had a fan belt break will appreciate the cam-driven water pump of the LT1/LT4 and LS1 engines, which allows the engine to be safely operated without a belt at all, giving you the chance to drive to a service station for a replacement engine accessory belt rather than having to walk to one. Note, however, that your alternator won't be turning, so the engine will draw its electricity from the battery and will eventually discharge the battery enough that the car won't continue to run.

The first-generation V-8 water pumps turned in opposite directions, depending on whether the pump was driven by a V-belt or a serpentine belt. V-belts rotated the water pump clockwise, while serpentine belts spin the water pump counterclockwise. This difference means that the water pumps are not interchangeable; the difference is more than just the pulley bolted to the front of the pump.

Whichever direction the pump spins, its efficiency can be greatly increased by installing a solid flow-director disk to the impeller. While this can be a do-it-yourself job (a sheet metal disk the same diameter as the impeller can simply be riveted to the impeller), many aftermarket water pumps feature that modification, along with heavier duty bearings, and even a lightweight, heat-shedding aluminum housing. And if your water pump has any miles on it, it's probably ready for replacement, so you might as well just replace it with a high-performance pump.

Thermostat

The thermostat controls the minimum coolant temperature once the engine is thoroughly warmed up. A temperature-

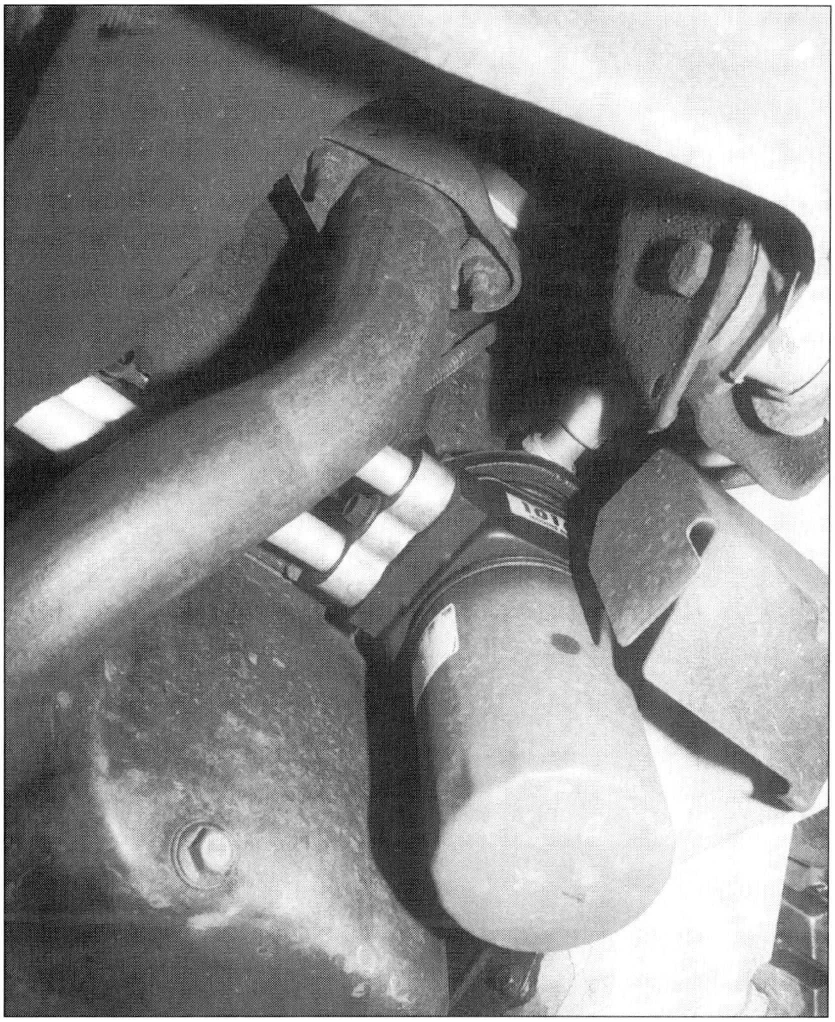

Here you can see the production oil cooler used on many Corvettes, Camaros, and Firebirds from the factory. An adapter mounts between the filter and the block. Coolant supply lines connect to the adapter. As coolant flows through the adapter, it picks up heat from the oil that flows through the adapter. It's a simple system that's easy to install, though aftermarket systems do often provide better cooling ability.

sensitive spring will shrink when coolant of a specific temperature hits it, thus opening a port through which coolant can flow out of the engine. Different thermostat temperature ranges use different springs to achieve those temperature settings. Essentially, a thermostat allows the engine's cooling system to respond to the engine's heat requirements by allowing it to maintain a given coolant temperature: When the coolant is cold, the thermostat stays closed, trapping coolant in the engine where it heats up more quickly than if it was permitted to flow freely through the engine. When the engine is thoroughly warmed up, the thermostat opens to permit the coolant to flow to the radiator to shed the heat it picked up in the engine, before returning back to the engine.

EFI engines—whether TBI or TPI—are particularly temperature sensitive, so you should never attempt to run the engine without a thermostat installed. As we discussed earlier, the thermostat itself provides a needed restriction to coolant flow, to prevent it from overcooling the engine as well as to shorten engine warm-up, especially in cold weather. Conversely, running without a thermostat may cause coolant to flow through the radiator too quickly to shed the heat it picked up in the engine, so the coolant never cools, and thus can absorb relatively little heat from the engine, so the engine actually ends up overheating. Either way, the end result is a problem, so don't replace a thermostat with simple flow-restrictor disks, which simply try to control flow velocity by forcing coolant through a disk with a hole in it. Some engine builders, however, point out that thermostats do induce thermal shock when they first open, because the coolant outside of the engine is cold, while the engine itself is hot, so when the thermostat opens to allow the hot coolant inside it out, the cold coolant flows in and contacts the hot engine and cylinder heads, "shocking" them, much as you would be shocked by jumping into a cold ocean in the middle of the summer. This can lead to distortion of the block and heads. Flow restrictor disks allow all the coolant to gradually warm up, however, nearly eliminating the chance of thermal shock.

Because the ECM factors data from a coolant temperature sensor when calculating fuel mixture, you need a thermostat that will allow enough heat in the engine to prevent the EFI system from operating in the warm-up (open-loop) mode, but not so hot that it becomes prone to detonation. In the old days, a 160-degree Fahrenheit thermostat was the ticket for V-8 engines, but at 160 degrees the EFI system will think the engine isn't quite warmed up yet. With a 170-degree thermostat, on the other hand, the engine runs only marginally hotter, but will prevent an extended, mixture-richening warm-up mode operation.

Hoses

The cooling system hoses—both the upper and lower radiator hoses, as well as heater hoses—represent some of the weakest spots in your engine's cooling system. Over time the hoses can become brittle and crack, releasing coolant. Or they could become so weak that, rather than cracking, they simply burst on the spot, spewing as much coolant as your engine can pump out.

From the factory, TPI engines are equipped with rubber cooling system hoses, and these should be inspected regularly for cracks and replaced approximately every three years or so—much more often for competition engines.

The important thing to notice in this photo is the factory power steering cooler line (arrow). As with your engine, your power steering system depends on its oil to prevent damage, as well as to provide the steering assist. A simple line like this can decrease power steering fluid temperature considerably and is easy to fabricate.

An upgrade for rubber hoses is silicone hoses. They are stronger and tend to last longer than rubber hoses and can prove quite reliable, even in competition use.

Even better than silicone hoses are braided steel hoses, which are exceptionally strong and can withstand not only higher pressures, but also higher flow velocities and higher temperatures. They're also less likely to be punctured in an accident. Just don't confuse braided steel hose covers for actual braided steel hoses. Braided hose covers are almost purely for cosmetic purposes. (Braided hose covers do provide modest heat shielding and puncture protection, but they can also chaff the rubber hose within the cover.)

Finally, it is possible to replace the bulk of the hoses under your TPI-powered car's hood with rigid tubing, using only short sections of hose to connect the rigid tubes to the radiator, water pump, thermostat gooseneck, and so on. Rigid tubes can also make quite a fashion statement if anodized or powder coated.

In any event, flexible "one-hose-fits-all" discount hoses should be avoided, because they disrupt coolant flow and also tend to fail more quickly where the hoses bend. Quality molded hoses tend to be much more reliable.

Coolant

Coolant is the second most commonly thought of component of your car's cooling system. The coolant is the blood of the system, traveling through the engine, picking up heat as it goes, then going to the radiator to dissipate that heat, before returning to the engine to pick up more heat.

The factory fills cooling systems for TPI engines with a near 50/50 mixture of antifreeze and water. Unlike many things that are mixed together, a mix of antifreeze and water actually lowers the mixture's freezing point, while raising its boiling point at the same time. The usually ideal 50/50 mix will typically freeze at around -30 degrees Fahrenheit and will boil at roughly 260 degrees Fahrenheit. Straight water freezes at a higher 32 degrees Fahrenheit and boils at a lower 212 degrees Fahrenheit.

If you happen to live in a warm climate, you may be able to mix water with a product called Water Wetter, which helps improve the ability of water to absorb and dissipate heat, plus raises the boiling point of the water. In colder climates, use of this product could allow the coolant to freeze in the engine, which could cause cracks in the block or cylinder heads, destroying them.

Electric Water Pumps

Anyone who's flipped through performance parts catalogs has probably noticed ads for electrically driven water pump systems offered by companies such as Moroso Performance. These systems are intended to provide necessary cooling ability without the power-robbing effects of an engine-driven water pump. A small electric motor draws power from the car's battery or electrical system, and, through a drive belt, turns the water pump's input shaft.

Unfortunately, because the electric motor is a constant-velocity device it isn't able to vary speeds (and thus the volume of coolant flowing through the engine) to correspond with varying engine speeds. Nor are the systems designed for extended use. They are primarily intended to be run by drag racers, who typically only need them for short periods of engine operation, including the first few minutes of time in the pit area between runs. The electric motors, however, are not heavy-duty enough to withstand continued use, such as would be needed for even a short drive around town. For this reason, in particular, electrically driven water pumps are not recommended for anything but drag racing applications.

Another easily overlooked part of the cooling system is the assortment of core plugs (aka, "freeze" plugs) that fill the various holes in the block and heads leftover from the casting process. Standard core plugs are steel, but high-performance replacements are brass, which tend to not only seal better but don't rust.

Note that until 1995, Chevrolet used ethylene-glycol antifreeze mixed with water. The mixture was a fluorescent lime-green color and needed to be changed every two years or so to maintain optimal efficiency. Beginning in 1995, a new, orange-colored antifreeze was used, referred to as DexCool, which is designed to last for up to 100,000 miles before needing to be changed. The two types of coolant cannot be mixed or internal engine damage may result.

In terms of cooling ability, regular ethylene-glycol coolant, DexCool, and even Water Wetter are roughly equal. DexCool should last longer, and Water Wetter doesn't offer any antifreeze protection, so, again, it's not suited to colder climates. Also, some people say you should only used distilled water, not tap water. In practical terms, however, there are no distinct advantages to running distilled water, except perhaps for slightly fewer deposits forming in the water over time. But if you change your coolant regularly and flush the system thoroughly at that time, deposit build-up shouldn't prove to be a problem anyway.

Fans

Radiators need to have air flowing through them to cool the coolant flowing inside their tubes; however, a number of factors can limit the amount of air that actually reaches a radiator, thereby minimizing its cooling abilities. For example, late-model Camaros and Corvettes don't actually draw much—if any—air from the front of the vehicle; instead, they draw air from beneath the car. On the other hand, trucks and cars such as the Impala SS do draw air through the grille straight toward the radiator, but they typically have thick air conditioning system condensers mounted to the face of the radiator, which also minimizes how much cool air reaches the radiator. In any event, the amount of cool air that naturally reaches the radiator may not be sufficient to dissipate the heat from the coolant. Fans alleviate this problem by pulling extra air through the radiator.

Most passenger car TPI engines have relied on electric fans mounted directly to the radiator, and those fans typically work very well . . . when they're on. And that's the catch: The stock settings for the fans are often too high for performance use, allowing the engine to build and retain too much heat. Fortunately, a replacement thermostat control switch from an aftermarket performance parts company will turn on the fans at lower temperatures, preventing the engine from getting hot in the first place. Likewise, most aftermarket PROM chips and EPROM reprogramming comput-

Machine shops can install freeze plugs for you for a minimal fee, or if you're feeling up to it, you can do the job yourself. You can tap them in with a socket the size of the freeze plug. Apply a little sealant to the plug before installing to ensure a good seal.

ers are programmed to activate the fans at lower coolant temperatures.

Trucks, however, typically still feature engine-turned fans with either thermostatic or speed-sensitive hubs. Thermostatic hubs engage the fan when the air flowing through the radiator is warm, which is typically at lower speeds. Speed-sensitive fan clutches rely on a viscous fluid in the hub to cause the hub's input shaft to rotate the hub and fan. At lower input shaft (engine) speeds, the fluid causes the hub to spin, in much the same way a torque converter works, but at higher speeds the shaft spins through the fluid, allowing the fan to free-wheel when vehicle speed is generally sufficient to produce enough airflow through the radiator to allow it to cool efficiently.

Secondary Cooling Systems

Anything that indirectly helps cool the engine or its auxiliary equipment is a secondary cooling system. While improvements to these secondary systems may not have as dramatic an effect on engine coolant temperature, they can have a considerable effect on performance or other engine characteristics. At the very least, paying careful attention to each will pay off in extended engine life.

Oiling System

Next to lubricating your engine's multitude of moving parts, helping to cool parts is your oil's most important function. If you stop and think about it, many parts in your engine never have coolant come anywhere near them. Bearings, the crankshaft, connecting rods, pistons, piston rings, cam drive systems, the camshaft, rocker arms—none of these parts are cooled (not directly, anyway) by engine coolant, each relies on engine oil to absorb heat and carry it away.

Of course, your oil can't absorb much heat if it's already overheating. Overheated oil can happen because of an insufficient quantity of oil in the engine; because of insufficient flow through the engine, perhaps due to a restriction somewhere; because of excessive engine temperatures (due to a malfunctioning cooling system or extreme operating conditions); or, of course, due to insufficient cooling of the oil.

Each of these problems usually has a fairly simple solution.

Insufficient quantity could be because the engine is simply low on oil. Oil may have leaked out, or perhaps the engine is burning oil. Your engine may also run out of oil if it is operated at high rpm for extended periods of time, especially if you are running a high-volume oil pump. Since oil is under pressure when flowing out of the oil pan, but returns by gravity, it's easy to understand how, at high rpm, the oil pump may empty the oil pan faster than the oil returns to it. A higher capacity oil pan could be one solution: By simply increasing the quantity of oil available, you decrease the chance all of it will be pumped out of the pan. You could switch to a standard-volume oil pump, which actually provides ample pressure and volume (with close-to-stock bearing clearances) as proven by the countless small-blocks that endure hundreds of thousands of miles with no oil pressure problems.

Professional racers, such as those on the NASCAR or SCCA Trans Am circuits, rely on dry sump oiling systems that actually draw oil from the engine after it is circulated, to ensure oil doesn't linger in the engine any longer than it needs to. (Dry sump systems typically have a storage tank that holds gallons of oil, too, as compared to the 4 quarts held by a standard small-block oil pan.)

Of course, no matter how much oil you have, you can still get it too hot. In those cases, you need to implement methods of cooling the oil. First and foremost are dedicated oil coolers. Two basic types are available: oil-to-coolant and oil-to-air.

The factory installed oil-to-coolant oil coolers on many production Corvettes and Camaros; these systems route engine coolant to a special adapter that mounts between the engine block's oil filter mounting pad and the filter itself. These coolers help oil reach its proper operating temperature quickly (because the coolant heats up quickly) and help maintain that temperature by shedding excess heat to the coolant. If you happen to run into a situation where the coolant is overheating, however, it can cause your oil to run hotter than is desirable, which can worsen overheating conditions by thinning the oil. Conversely, if the oil is running hot, it can cause the coolant flowing through the adapter to pick up additional heat.

Oil-to-air coolers, on the other hand, work similarly to your cooling system's radiator. An adapter between the block and filter routes oil to the mini-radiator (usually positioned in an area of high airflow). Heat transfers from the oil to the cooler's tubes, then to its fins, then to the air flowing through the cooler. These coolers are the type typically used in competition engines, where the complexity of an oil-to-coolant system would pose an unnecessary risk of a coolant leak or where high coolant temperatures may cause the oil to heat up more than would be desired. Oil-to-air cooling systems are more easily adapted to various vehicles as well.

You can also improve cooling by simply using smarter replacement parts for some of your oil system's existing parts. Aluminum oil pans, for instance, dissipate heat more readily than stamped steel pans. Likewise, rocker (valve) covers with built-in oil drippers—or better yet, oil sprayers—can increase oiling of the valve springs, which reduces their temperature and thus reduces fatigue, which could lead to a broken valve spring in high-rpm competition. This does raise the oil temperature, but a well-designed oil cooling system can more than compensate for the modest temperature increase.

Of course, the oil you use also has a great deal to do with your oil system's ability to help cool—and protect—your engine. Cooling-wise, a high-viscosity oil (such as 15w50) is thick, and thus flows less rapidly through the engine, so it reaches its maximum temperature quickly and can no longer absorb heat from the engine at that point, so the engine will run hotter. Thinner oil, such as 5w30, flows more easily, so it can efficiently pick up heat in the engine and carry it away, thereby allowing the engine to run cooler. In severe-use conditions, oil quality becomes an issue, because conventional oils will begin to burn and vaporize, which leaves less oil to absorb the engine's heat, so engine runs hotter. In addition, the oil forms a thick, tarlike substance that gums up engine parts and clogs oil passages, further limiting how much oil can flow to different areas of the engine to absorb heat.

While any high-quality conventional oil will work well when changed regularly (every 3,000 miles or six months, whichever comes first), synthetic oils provide added insurance. Synthetic oils are engineered to withstand higher temperatures than conventional oils, so they don't thin out as quickly when they get hot, and they can get hotter before burning off. An old Mobil One television ad features two frying pans—one with conventional oil and the other with Mobil One synthetic—subjected to high heat. It clearly demonstrates conventional oil burning, turning to a thick, tarlike substance, while the Mobil One was still golden brown and flowed smoothly. Advertising theatrics? Probably, but the premise is sound, and if you require proof, just look at what oils professional motorsports teams use in their vehicles.

Synthetic oils can withstand heat so well, in fact, that while Chevrolet installs oil-to-coolant oil coolers on Camaros and Firebirds (which leave the factory with conventional motor oils), Corvettes leave the Bowling Green, Kentucky, assembly plant with Mobil One in their oil pans, but without oil coolers!

Induction System

Stretching the concept of cooling systems a bit further, we can include the issue of providing cool, dense air for the combustion process. A cooler intake charge helps cool the combustion chamber, so the heads and pistons run cooler. The problem is, there are lots of things working against you when it comes to delivering cool air to the engine. Underhood temperatures on high-performance vehicles can be staggeringly high, plus the very designs of some of the systems and components you bolt on your engine can accentuate any problems. Also, cooler combustion chambers and pistons lower the combustion temperatures somewhat, so some power is lost as a result of the cool intake charge, though that amount is basically negligible.

Because underhood air is typically so hot, the best way to maximize the air being drawn into your engine is to draw it from outside the vehicle with a cold-air induction or ram-air kit. These generally consist of a scoop of some sort, some form of ducting, and a high-flow air filter to prevent debris from entering the engine. Some systems employ a vehicle-specific scooped hood, while others are more universal, consisting of a somewhat generic scoop that mounts under a bumper or behind a grille.

Once you've got nice, cool air flowing toward your engine, you need to ensure that it stays nice and cool. That involves minimizing the temperature of parts with which the air comes in contact, specifically the throttle body and intake manifold assembly.

The throttle body is heated both to prevent it from icing up in cold weather, as well as to aid in warming the engine more quickly. If ice forms in the throttle body venturis, the throttle blades will stick, which will eliminate your control of the throttle. In warm weather, however, the air charge is already warm enough to not pose a risk of freezing in the venturis, but the throttle body's coolant passages heat the already warm intake air charge, reducing the air's density and diminishing the quantity of fuel the air can absorb, and thus the peak power the engine can produce.

Your intake manifold also tends to heat intake air, because the manifold itself is heated by both the coolant that flows through it (on first-generation engines) and by the oil flung off the camshaft against the bottom of the intake. Non-TPI engines also face an additional source of heat: the heat crossover passage that directs a portion of the exhaust gases from the center exhaust ports through the intake manifold to aid in engine warm-up and to prevent manifold icing. The crossover is only useful for wet-style intakes—those that flow both air and atomized fuel—because an excessively cold intake could cause the vaporized fuel to precipitate out of the air and form droplets.

To solve the problem of oil splashing on the bottom of the intake, you can install an oil splash shield, which reduces the chance that hot engine oil will heat the intake manifold. Sheet metal shields are available or can be fabricated to block oil from contacting the intake. A splash shield is especially important for nonroller cam engine blocks, since they lack the sheet metal rocker-retaining spider, which provides some shielding for the intake from hot oil flung off the cam.

As for the heat crossover, you can either minimize its effect or completely eliminate it in a number of ways. First, most intake manifold gasket sets contain heat crossover passage restrictors, which can be used to reduce the size of the passage, but still allow some exhaust air to flow through the intake to afford better cold-weather driving. If your goal is to block the crossover entirely, you can install gaskets from the major gasket producers, such as Felpro, that do just that. Or if you're swapping intake manifolds or cylinder heads, you can select pieces that lack the crossover passages.

A last note about induction system cooling: Drag racers are familiar with the trick of icing down the intake manifold between runs to minimize its temperature. As crude as this trick is, it does work—briefly. The key is to keep the engine off until the last possible second, because as soon as you start the engine, the heating process begins.

Fuel System

In the old days, it wasn't uncommon for an engine to suffer vapor lock—when the fuel literally boils in the fuel line, creating a gas pocket that blocks fuel from reaching the carburetor. Since EFI engines all feature electric fuel pumps, that push rather than pull, vapor lock is virtually unheard of these days.

Another aspect of EFI fuel systems that minimizes the chance of vapor lock, and actually aids performance, is the employment of fuel return lines. Rather than a single line that runs only from the fuel tank to the fuel injectors, EFI systems use a supply line plus a fuel return line that sends unused fuel back to the fuel tank continually. Because the fuel never sits still near a heat source, it stays pretty cool.

It is still possible to improve the system using another old racer's trick, however. That trick is a fuel line "cool can" canister through which you run the fuel line and pack with ice to cool the fuel within the line. You can either make a cool can using nearly any canister (a large coffee can works well, or even a large plastic jug), and a coil of fuel line. Just remember that EFI fuel systems are under high pressure and should only be used with solid steel or stainless steel lines.

While the stock electric fans with which the factory outfits TPI-powered cars flow more than enough air for even high-performance and racing use, the one problem they do face is when they activate. Stock systems rely on signals from the ECM to turn them on. A new PROM chip or an EEPROM reprogrammer unit will typically lower the temperatures at which the fans turn on. Alternatively, a device like HyperTech's Cool Fan Switch alters the signal voltage that reaches the ECM, causing even a stock ECM to activate the fans earlier. *HyperTech*

Ignition System

Your ignition system isn't so much a cooling aid but rather something that affects the engine's ability to produce heat (power), and therefore affects how hot or cool the engine runs. Your ignition system's efficiency and timing adjustments are both important factors that influence your engine's heat management. If your ignition system doesn't thoroughly burn the intake mixture, the amount of heat (i.e., power) produced will be less than that produced with a healthy, efficient ignition system.

Similarly, if your ignition system has to be run with lots of advance, to compensate for high compression, poor fuel, or other detonation-inducing factors, the combustion process will start well before TDC (top dead center), so cylinder pressures are reduced, as is power (heat) production. Thus, an engine running lots of ignition advance should run cooler than a similar engine run with less advance.

Of course, as we've explained, both of these scenarios result in reduced power output, so neither is desirable. Interestingly, however, power reduction is the method that GM employs in the Gen-III engines (LS1 and 1999 Vortec V-8s) to allow them to run without any coolant. The engine's control system kills the ignition pulses to cylinders on one bank, so the air pumped in and out of those cylinders is not burned and thus produces no heat and actually absorbs some heat from the engine, allowing it to run cool enough that coolant isn't necessary.

For practical purposes, unless you're experiencing problems with your engine running hot, there's little reason to purposely alter your ignition timing to change its operating temperature. The power you'll give up is far greater than the benefit you'll derive from the modest heat reduction. In other words, there are better ways to control overheating problems than by altering ignition timing or operation.

Accessory Systems

A few other engine components need some help shedding—or fighting off—heat, if they are to work reliably and properly.

Chevrolet small-blocks are notorious for starting difficulties when thoroughly warmed up. The problem (generally) stems from exhaust manifold or header heat that causes the starter-mounted solenoid to temporarily fail. There are several possible cures for this problem. The easiest cure is to purchase or fabricate a simple sheet metal heat shield that you position around the solenoid, but with a bit of an air gap between it and the solenoid to allow the solenoid to dissipate heat.

Late-model and aftermarket high-torque gear-reduction starters are designed to withstand the heat of cramped engine compartments, and so they generally do not experience the hard starting conditions. Gear-reduction starters also have the added benefit of being able to deliver higher cranking rpm for faster starts, as well as higher-torque cranking for high-compression engines. As if that weren't enough, they also tend to be significantly smaller and lighter than standard starter motors.

A third option does exist: You can modify the starter to use a remote-mounted solenoid, similar to Ford starter systems; however, the gear-reduction replacement starter motors are a better solution and far easier to accomplish.

Another accessory that experiences a fair amount of heat that needs to be dealt with is the power steering pump. When run for extended periods of time, or worked very hard, such as during autocross runs, power steering fluid heats up considerably. The solution is to install a power steering fluid cooler. A few options exist here. First, General Motors equipped some vehicles with power steering coolers that were nothing more than lengths of tubing mounted in an area where they will receive moderate airflow. The fluid simply flows through the tube, and air blowing across the tube whisks away heat.

A similar sort of system was employed for many years on police cars build by GM, Ford, and Chrysler. This system mounted a miniature radiator about the size of a thick candy bar off the power steering pump assembly. A tube running through the fins of the radiator carried power steering fluid, and air flowing through the fins dissipated the heat. Simple and effective. Such units are often available in salvage yards or can sometimes be ordered new through dealership parts departments or even auto parts chain stores.

C O N T E N T S

CHAPTER NINE

NITROUS OXIDE, SUPERCHARGERS, AND TURBOCHARGERS

Internal combustion engines are merely air pumps: They draw it in (mixing it with fuel in the process), burn it, then blow it out. The better your engine is at doing this, the more power it can make. That's why engine builders spend so much time working on free-flowing intake manifolds, cylinder head ports, and exhaust systems.

But if you're not so handy with a grinder, or if you've already done all that and are still looking for more power, you may want to consider artificial means of increasing your engine's breathing efficiency.

Nitrous oxide systems, superchargers, and turbochargers all make even a stock engine act like a purpose-built race engine; and they transform race engines into better, more powerful race engines.

What Do They Do?

Don't blame us, but no matter how good you are at tuning your Chevy's naturally aspirated engine, it's still pathetically inefficient.

Sure, the amount of emissions gases given off may be minuscule, but so is the power it generates, compared to the potential afforded by the engine's cavernous cylinders.

The problem is a simple law of physics: Differing pressures always seek to equalize. Every time a piston travels down a cylinder on an intake stroke, it creates a vacuum (technically it's just a low-pressure zone, not really a vacuum). Since that zone is lower than normal atmospheric pressure (around 14 pounds per square inch at sea level), the atmosphere (air) tends to rush into the engine to fill the low-pressure zone, much the way water rushes to fill the hole made by a rock dropped into a puddle.

But there are two problems with this: First, is that the engine doesn't leave the atmosphere much time to fill that low-pressure zone. Remember that the camshaft will snap the intake valve closed after only a fraction of a second. So the pressure inside the engine may or may not be able to equalize with the surrounding atmosphere.

The second problem is that even if the pressure does equalize, the cylinders still aren't nearing their potential for power production because they actually contain relatively little air and fuel.

Fortunately, there are a few ways to increase an engine's ability to produce power, and they're nothing new to the automotive world. Nitrous oxide kits, turbochargers, and superchargers all increase an engine's power output by increasing the volume of oxygen and fuel an engine can inhale. Interestingly, though, each does this in a somewhat different manner and with differing advantages and disadvantages.

Despite the common goal, each of the three types of systems operate on the same basic premise: They get more oxygen and fuel into your engine so that

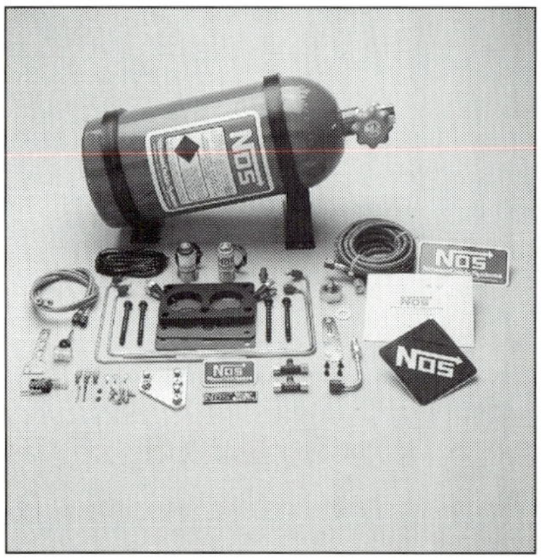

Horsepower in a bottle. Laughing gas. Nitrous oxide is called lots of different things, but no matter what you call it, call it effective. By injecting nitrous into the intake charge, nitrous cools the air, making it more dense and thus able to hold more fuel. It also promotes a better burn in the combustion chamber. All of which results in increased power literally at the touch of a button. This is Nitrous Oxide Systems' kit for TPI engines, and it comes complete with everything needed for an afternoon installation. *Nitrous Oxide Systems*

the combustion process produces more heat, and thus more power.

The bottom line, however, is that each of these systems can transform a ho-hum engine into a blazingly fast rocket. In many cases, nitrous, superchargers, and turbos can be directly bolted onto a stone-stock Chevy TPI engine to boost its power output by 100 or more horsepower, resulting in even better performance than an expensive engine that was purpose-built to deliver a given power output.

Read along as we explore the pros and cons of so-called "power adders" that force your engine to make more power.

Nitrous Oxide

Nitrous oxide is a popular and effective power-booster for stock and race engines alike. Nitrous oxide (also known as laughing gas, because of the effect it has on people if inhaled—which you should never do without medical supervision) increases engine power in two ways: First, it super-oxygenates the air-fuel mixture, which allows more fuel to bond with the air, so the combustion process produces more heat, which translates into more power. Second, nitrous oxide tends to cool the air/fuel mixture, which both allows the air/fuel mixture to carry even more fuel, thanks to the increased density and oxygen content, plus also helps to reduce hot spots in the cylinder that could lead to detonation. Since the intake air density can support so much more fuel, nitrous kits inject more fuel to prevent the engine from running lean, which could cause it to burn a piston or detonate so severely that it could damage the engine.

Easy-to-install bolt-on nitrous kits are available for nearly all EFI Chevrolet engines from a number of manufacturers and can generally boost power anywhere from 75 to 150 horsepower for a stock engine. Modified engines can often handle additional power boosts, up to around 250 horsepower, while full-race motors with heavy-duty reciprocating assemblies may be able to withstand as much as 400 nitrous-generated horsepower or more!

TBI-equipped engines can usually use a normal injector plate that gets sandwiched between the intake manifold and the throttle body assembly, while

Ahhhh, superchargers—the one thing better than more cubic inches. These are centrifugal superchargers from Vortech. A belt driven by the crankshaft spins the impeller inside the housing, causing it to ingest air through that rather large hole in the middle of the unit, then blows the air around the housing and out the outlet port, where it then runs to the throttle body or carburetor. *Vortech*

TPI/LTx cars typically use a plate that mounts between the throttle body assembly and the intake plenum. Note that this means that fuel is now suspended in the air that flows through the intake manifold, and that can lead to uneven fuel mixtures cylinder-to-cylinder or even puddling in the manifold. More exotic systems solve this problem by using injectors that feed directly into the intake manifold runners, aimed toward the intake valves.

Nitrous kits are easy, cost-effective, and safe methods of increasing engine output for short periods of time, such as is needed during a drag race; however, you have to be cautious about how much extra power you attempt to make with a nitrous kit, because it is possible to overstress parts and cause them to fail, or to at least increase wear. When in doubt, it is best to be conservative with nitrous system jetting and careful in its use, limiting nitrous use to 15 seconds or less and only at full throttle.

Additionally, it is often beneficial to employ timing retard devices in conjunction with nitrous oxide systems. Because nitrous oxide–assisted engines create higher cylinder pressures, ignition timing should be retarded. But since you only use nitrous sporadically (and never on the street!) you don't want to hamper engine performance during the 99 percent of the time that you're not running nitrous through the engine, so ignition accessories systems that can be wired to

Here's a Vortech supercharger as installed on an LT1 Gen-II engine. Note the compact design. The Vortech kits are engineered to fit not only particular engines but particular vehicles' engine compartment confines for easy installation that doesn't require any fabrication. *Vortech*

activate when the nitrous activates retard the timing only when needed, allowing engine timing to be optimized for whatever kind of driving you're doing.

Nitrous Benefits

Based on all that, you should be able to spot the good points of running nitrous: It's inexpensive (initially, at least); it's easy to install; it doesn't affect around-town driving; and some systems are even tunable to deliver more power as your needs increase.

Serious drag racers may like the fact that it's possible to hook up nitrous kits in stages too. So you can have a relatively modest kit for launch, when cylinder pressures are greatest as the engine strains to get the car moving. Then, when the car is moving, a higher-power kit can kick in to boost power even more for continued acceleration. These systems aren't exactly common on TPI engines (and they're virtually unheard of for TBI systems), but if you're building an all-out EFI drag car, staged nitrous kits may be just the ticket.

Nitrous Drawbacks

Since we all know there's no such thing as a free lunch, it should come as little surprise that nitrous systems do have a few drawbacks that warrant at least some consideration on your part.

First, from an economic standpoint, it costs money to refill those nitrous bottles. Sure, the system is cheap to install, but if you use it regularly, you're going to find that refilling the bottle will significantly drain your wallet.

Another Vortech kit, as installed in a TPI first-generation Camaro's engine compartment. Note the clean installation. Because the system feeds directly to the throttle body, air is filtered prior to entering the supercharger unit. *Vortech*

Ram Air Induction

While the benefits of supercharging and turbocharging have been exploited for more than half a century, both systems were expensive and difficult to implement. Ironically, it wasn't until the 1960s that the Big Three auto makers realized they could get a free form of forced induction by taking advantage of a car's movement through the air to force air into the engine.

In practice, the systems were known as ram air systems, and it worked like this: Scoops, usually mounted in the hood, behind the grille or even under the bumper, fed air into hoses that routed the air to a modified air cleaner assembly. The faster a car traveled, the more air flowed through the scoops and hoses to the engine.

The systems didn't achieve anywhere near the power levels of superchargers or turbos, but they were inexpensive and were effective. And if nothing else, they gave the auto industry something to hype in advertising.

The fact that your supply of nitrous dwindles with each use also means that you could run out at the least opportune time—either between rounds at the track, or worse, in the middle of a run. At the very least, you're likely to lose races that way, but there's also the possibility of engine or drivetrain damage if the nitrous starts sputtering in midrun.

Beyond that, nitrous isn't well-suited to street use, because it creates a sudden explosion of power that can make a car difficult to control on the street, around other cars or people. So rarely will you encounter conditions on the street in which nitrous would be of benefit.

Then there is always the issue of whether nitrous oxide will increase wear in your engine, or cause something to break. In truth, that is possible, but it shouldn't be a problem if you properly match the nitrous system to your engine's abilities. In other words, stock engines—especially ones with lots of miles on them—won't be able to withstand the increased cylinder pressures and forces created by high-horsepower nitrous kits, but a low-power kit should work just fine.

Forced Induction Systems

Stock Chevrolet engines—even EFI engines—are normally aspirated, meaning the only fuel and air they can get into their cylinders is however much they can suck in. But by pressurizing the induction system, it is possible to force-feed the engine more air and fuel than it would normally get. With more air and fuel in the cylinders, the combustion process will generate more heat, and as a result of that heat, the burning gases would expand with greater force, creating more power.

There are two basic kinds of forced induction systems: turbochargers and superchargers. Each has its own set of particular strengths and weaknesses, and use of them should be considered carefully because they place added stress on engine assemblies and could cause failures. Additionally, each generally requires specific combinations of parts to be most effective, which could require you to rebuild your entire engine.

Turbochargers

Turbochargers use exhaust flow to turn a turbine. As the turbine spins, the shaft to which it is connected spins. On the opposite end of the shaft is an impeller, which, naturally, spins as the shaft spins. This impeller blows additional air into the engine. As more air enters the engine, more fuel is mixed with it, and the cylinders are completely filled by the pressurized induction system with fresh air and fuel (the pressurization also helps to evacuate exhaust gases), so the combustion process produces higher-than-usual levels of power.

Because turbochargers are driven by the flow of exhaust gases, however, they require the engine to reach a certain rpm before the gases turn the turbo fast enough for the boost developed to be of value. In other words, at low engine rpm, the exhaust gas velocity doesn't drive the impeller fast enough to create boost, so the engine acts as a naturally aspirated engine. But as the engine speed increases, so does that of the exhaust gases, and thus the turbo impeller. The time it takes for the impeller to speed-up to a point at which it begins producing boost is felt and known as turbo lag. Another problem that stems from the use of exhaust gases to run the turbo is excessive heat.

The heat from the exhaust heats the turbo unit, which causes the air flowing through it to the intake manifold to heat up, which makes the air less dense. Thus there are fewer oxygen molecules for fuel to bond to, so fuel content (per volume) is reduced, and—by extension—so is power. This problem is compounded by the fact that as you pressurize the air, it naturally heats up, and the problem escalates as the turbo continues to run since exhaust gases are hotter from the pressurized combustion cycles.

Use of an intercooler helps combat the heat problems insofar as it lowers the intake air temperature substantially and thus increases its density. An intercooler is simply a radiator, through which the intake air is circulated to cool it, much as coolant is chilled by circulating through the cooling system's radiator. So the airflow path is as follows: Air enters the air filter housing (or possibly a duct leading to the air filter housing), then flows through ducts to the turbo unit, which compresses the air and sends it to the intercooler, then it flows to the throttle body assembly.

It should also be noted that, because the turbo units get so hot, they are extremely hard on engine oil that is supplied to the turbo. In fact, the number one cause of turbo turbine failure is improper lubrication, usually due to oil that has thinned out due to the extreme heat. Synthetic oils are highly recommended for turbocharged applications.

Also, as with nitrous oxide kits, turbochargers require retarded ignition timing because of the increased cylinder pressures, and they also require reduced compression to avoid detonation on pump fuels.

There are few off-the-shelf turbocharger kits for EFI Chevy engines these days, though it is possible to "home-brew" one.

Turbocharger Benefits

The good news is that turbochargers are about the closest thing to "free power" as you can get. They take a waste product (exhaust gas heat) and put it to good use to force more air into the engine. The process feeds itself, too: the more air the turbo forces in, the more exhaust blows out and at greater speeds, so it can blow even more air into the engine. These are important differences from superchargers, which actually use some of the engine's power to turn the impeller.

Turbos are also no-brainers as far as their use goes. If exhaust flow is high enough (that is, if engine speeds and cylinder pressures demand it) then the turbo provides boost, otherwise it simply freewheels. And while they aren't doing anything, they're not creating any unnecessary strain on the engine, so wear on the engine is minimized.

Turbocharger Drawbacks

Again, nothing's perfect, and turbochargers are no exception. One of the biggest problems you will confront is that there are few ready-to-install kits available. Those that are available rarely come cheap.

If those two problems don't scare you away from a turbo setup, then you'll need to prepare yourself for the not-too-simple process of setting up your car to work well with a turbo. Beyond simply installing the turbo system—which typically requires a low compression ratio to run reasonable boost—you will need to work on a chassis setup that will let you get your engine's power to the ground, which can be especially challenging, given that you will have to contend with turbo lag. The problem, in a nutshell, is that you've essentially got two engines: When the turbo isn't making boost, it's running naturally aspirated and making relatively little power; when it is making boost, you'll have an insane level of power. Finding a good chassis setup that balances these two extremes isn't an easy task. Time spent working with waste gates and careful selection of the size of

Nitrous: Is It Really the Best Bang for the Buck?

When comparing nitrous, superchargers, and turbochargers, one thing becomes clear almost immediately: There's a huge price difference between them, and nitrous—being the cheapest of the three choices—starts to look awfully attractive, as best it should, since it is an excellent technology.

But before you plunk down your cold hard cash, you need to consider the long-term ramifications of choosing nitrous versus other methods of boosting power. For a few hundred dollars, you can get a nitrous kit, and within a couple of hours have it installed and ready for use. This compares quite favorably to the thousands of dollars you would spend on a supercharger or turbocharger system and their respective hours or days of installation time. But with superchargers and turbochargers, that's where your costs essentially end. With nitrous, the cost of the kit is only the beginning.

Because nitrous relies on a depletable supply of nitrous-oxide gas, after a few dozen quarter-mile passes, your bottle's empty and your wild rides are over until you get a refill, at about $50 or more, depending on the size of your bottle and local refill rates. And that doesn't include the time it takes for you to actually go get the bottle refilled, which, it should be pointed out, can't be done at your local gas station.

If you start breaking down your cost per mile for a nitrous system, it can quite easily have the least desirable ratio. That's not to say you shouldn't add nitrous to your TPI-powered machine; nitrous may, in fact, be your best choice. But it pays to understand that you'll keep paying to run nitrous.

(In fairness, superchargers do periodically require a replacement belt, after several thousands of miles of operation.)

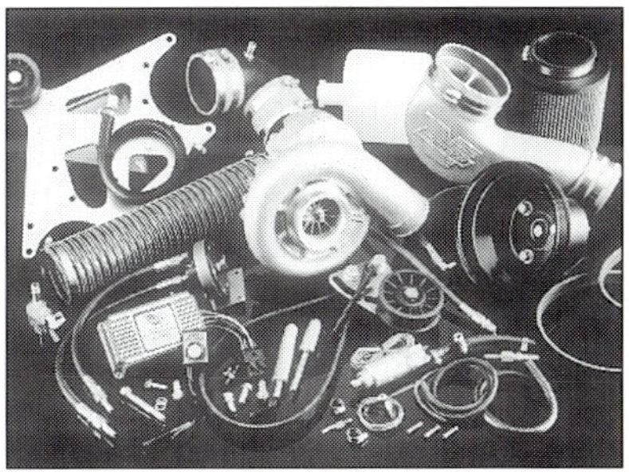

While all the hardware makes a supercharger kit, like this one from Vortech, look complicated, it's actually a rather straightforward installation. Instructions are provided, of course, to make sure you don't encounter any trouble. *Vortech*

Turbocharged engines are much less common. About the best known, however, are the Callaway Twin Turbo Corvettes of the middle to late 1980s. The large radiator-like devices on each side of the intake plenum and runners are air-to-air intercoolers to reduce the intake air temperature to hold more fuel for full efficiency. The turbo units, which actually resemble superchargers, are below the intercoolers, mounted to the exhaust manifolds for direct access to the exhaust flow that powers them.

Blowin' in the Wind

When choosing superchargers there are two popular types of systems: roots and centrifugal superchargers.

Roots superchargers are the type used on professional drag racing engines and are probably what most people think of when they first hear the word supercharger. They require specific intake manifolds, to which they are directly bolted. The carburetors or throttle bodies bolt to the top of the supercharger housing. A belt, driven by a pulley on the crankshaft spins the blower's input shaft and rotor, and a gear on the input shaft turns a similar gear on the driven rotor of the roots supercharger. The two rotors have a twisted, fluted design that forces air into the intake manifold in much the same way that an oil pump forces air into the engine. The rotors create a true vacuum in the blower housing that draws air through the carburetor or throttle body. Typically, roots-style superchargers will require hood scoops for additional underhood clearance.

Centrifugal superchargers are also driven by a belt; however, they use a turbocharger-like, centrally mounted impeller to force air through the housing, out an outlet port into ducts, and finally to the carburetor or throttle body.

For street use, the centrifugal blowers are easier to package within an engine compartment since they are mounted off the front of the engine, much like an alternator or air conditioning system compressor.

the turbo unit you run will make life easier, but it still won't be easy.

Once you do get the turbo set up and your chassis dialed in, there is the issue of heat buildup, both in your intake air and in your engine's oil.

An intercooler, either air-to-air or air-to-water, will help reduce the intake air charge temperature, so that it remains dense and thus able to hold more fuel. But intercoolers and the plumbing to them can be difficult to package in a tight engine compartment.

As for the oil, the solution is twofold: One, run synthetic oils, which are designed to withstand higher temperatures better (recall the Mobil 1 oil commercial with the conventional oil burning up in the frying pan, but Mobil 1 flowing freely); two, use an oil cooler to ensure the oil doesn't overheat during competition use.

Superchargers

While turbochargers are part-time power builders, superchargers run full-time and always keep the intake tract pressurized because they run off of the crankshaft via a drive belt. This extra drag on the engine does result in a minor parasitic power loss, but the benefits of a supercharger far outweigh that minor drawback. Otherwise, turbochargers and superchargers are essentially similar in function.

Because superchargers do not suffer from lag, as do turbos, it is easier to tune an engine for them, and it is easier to set up a chassis for one and to drive a supercharged vehicle, since power application is smooth and consistent, not sudden, bordering on uncontrollable.

Additionally, superchargers do not suffer the same heat problems caused by running off the exhaust gases, though they are still subject to the intake air temperature rising as a result of pressurization.

As with a turbocharger, using an intercooler will result in a fairly dramatic drop in intake air temperature, so there's more oxygen for fuel molecules to bond with and thus the engine makes more power. Use of synthetic engine oils is also beneficial for superchargers.

A number of companies offer supercharger kits for EFI Chevy engines, for both V-8 and V-6 applications; however, you should be aware that there are different types of superchargers available. See the sidebar "Blowin' in the Wind" for more information.

Superchargers also require retarded ignition timing and reduced compression, due to the increased cylinder pressures generated under boost. Off-the-shelf units are generally designed to deliver relatively little boost, because of the compression ratio of stock engines. If you purpose-build an engine to be supercharged, however, you can tailor the compression ratio and boost developed by the supercharger to deliver maximum power.

Supercharger Benefits

The allure of superchargers is that they create full-time power increases—at any engine speed, any throttle position, every driving situation. Since the supercharger is always providing boost, power application is always smooth, though somewhat amplified compared to a normally aspirated engine. This makes setting up a chassis much easier, and it also aids in drivability and controllability, since the engine's reaction to the movements of your right foot are predictable.

Since superchargers are not driven by hot exhaust gases, overheated air and oil aren't quite as problematic with superchargers as with turbos, though the use of an intercooler, synthetic oil and an oil cooler are still recommended.

Installation of a supercharger is somewhat more straightforward as well, since the units simply bolt-on, rather than requiring replacement or modifications to exhaust manifolds and systems. Placement of the system and the lack of a modified exhaust system means superchargers are generally easier to fit into an engine compartment.

In terms of upgradability, if you rebuild your engine so that it can withstand more power (stronger pistons and connecting rods, better pistons rings, and so on) then you can increase supercharger boost by simply changing the pulley ratio: Turn the supercharger faster and you get more boost and more power; slower speed yields less boost, but increased engine life.

In addition, it is possible to combine supercharging with nitrous oxide for an even greater increase in power for racing applications.

Supercharger Drawbacks

The number one drawback of superchargers is their price. Though a number of companies offer kits for Chevy EFI engines, they don't exactly give them away. It should be noted, though, that the kits are all-inclusive, right down to the hardware you will need to bolt them on, the drive belt, mounting brackets, and necessary ductwork.

Superchargers, however, like turbos and nitrous systems, do increase cylinder pressures and power, and that means that they increase the stresses and forces imparted on your engine's various components. Any weaknesses will quickly make themselves evident in the form of failures.

Though this is a somewhat subjective drawback, superchargers—and turbos, to an extent—are noisy. All that air rushing through them, plus the internal gears, creates a whining whistle that some people may find objectionable. To be entirely fair, the noise isn't overwhelming, and most enthusiasts would more than gladly sacrifice a little quiet for the impressive power gains.

As was mentioned before, superchargers do heat up the intake air because of the pressurization, but an intercooler sufficiently reduces the air charge temperature. Oil also gets heated up, but again, synthetic oil and an oil cooler dispense with that worry too.

Which One Is Right for You?

If you've been reading this book this far—or at least have been involved in performance cars for any length of time—you should already know that there is never a simple answer to that sort of question.

Depending on how your engine will be used and the parts and preparation that you are putting into it, any one of these solutions could work. Or any one of them could be disastrous.

The following information should be viewed as a set of very generalized guidelines, since it's impossible to account for every possible engine and chassis combination, driving conditions, driver skill, and more.

High-Performance Street Driving

Supercharging is usually your best bet for street use, because of the smooth, consistent, predictable delivery of power. Superchargers work well with either manual or automatic transmissions, and you could even run one pulley to produce modest boost on the street, then swap for a different pulley for more boost at the track.

Autocrossing

The extended runs of autocrossing, plus the widely varying throttle usage effectively eliminates nitrous from the running, while the varying engine speeds and throttle position alone knocks a turbo system out of the running.

Autocrossing needs a smooth application of power, so, if some form of power booster is desired, a supercharger presents the most logical choice; however, as always, you need to consider what installation of such equipment will do to your ability to compete in a particular class.

Drag Racing

Drag racers have long employed both nitrous oxide and superchargers—both individually and together—to help them get down the quarter-mile in a hurry. The short runs, steady throttle action, and controlled environment make drag racing ideally suited to the use of nitrous oxide.

Turbochargers would possibly be more popular in drag racing were it not for the difficulties created by turbo lag. So, since superchargers provide the same benefit without the lag problem, they're a favorite way of boosting a drag engine's output.

Road Racing

The demands—namely duration—of road racing rules out any meaningful use of nitrous oxide. Not to mention that most racing classes would forbid it.

Superchargers aren't particularly well suited to road racing use because of their drive belt arrangements, which don't generally withstand sustained high rpm.

Turbochargers, on the other hand, don't rely on complicated drive belt arrangements, so there is less chance that they will stop functioning because the belt slips, breaks, or gets thrown.

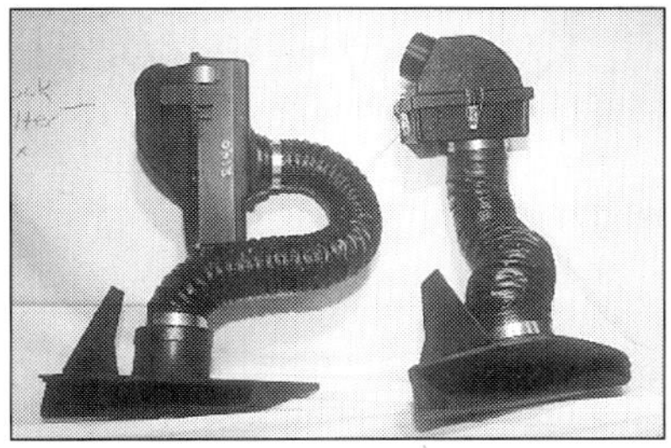

A cheap form of forced induction is a cold-air (or ram-air) induction system, such as this one from Lay Enterprises. The scoop is mounted beneath the front fascia to gulp fresh air as the engine drives through it. The scoop funnels the air through a hose to the air filter then to the engine. At low speeds the system provides cold, dense outside air for better efficiency. At higher speeds, the system provides a slight ram effect, similar to that of a supercharger though much less effective. This particular system uses a stock air filter box. *Lay Enterprises*

C O N T E N T S

CHAPTER TEN

CHEVY'S OTHER EFI ENGINES

THE LS1, VORTEC V-6S AND V-8S, AND TBI MOTORS

As the title of this book implies, the bulk of its material is based on and geared toward the Tuned Port Injection series of engines, including the 1985–1992 305 and 350 V-8s, as well as the LT1, LT4, and L99 Gen-II small-blocks, which ran from 1992 through 1996 (the 4.3-liter L99 died off in 1994).

While these engines are currently the engines of choice for performance applications, Chevrolet has produced a number of EFI engines in the last decade or so, most of which will serve you well as the foundation of a high-performance engine.

This chapter will briefly identify the different EFI engine families, their advantages and disadvantages, and summarize their suitability to performance applications.

The Gen-III Small-Block

Perhaps the ultimate production small-block Chevrolet V-8 engine, the third-generation 5.7-liter LS1 debuted beneath the hood of the 1997 Corvette to equal parts excitement and apprehension.

The first-generation small-block—which lived from 1955 all the way through 1997 virtually unchanged—had long since become a favorite with the hot rod crowd. The engine was small, light, and powerful, and there were scads of performance parts available to build it as mild or as wild as anyone could want. Quite simply, the small-block Chevy V-8 is the performance engine against which all others are compared.

But as good as the small-block was—even with the changes Chevy made with the Gen-II small-block, more commonly known by the names LT1, LT4, or L99—the handwriting was on the wall for the venerable old small-block. The old engine had simply reached the end of its useful life; it needed too many parts redesigned to eliminate inefficiencies, improve econ-

Chevy's first—and most prevalent—EFI systems are Throttle-Body Injection units, like the system used on the 1984 Corvette.

TBI systems were the fuel-metering and mixing systems of choice for GM's truck line from 1985 through 1995. This is a 1995 305-cubic inch V-8 with a TBI system.

The unit itself is pretty straightforward: an injector sits over each of the two venturis and sprays fuel into the venturi when signaled by the ECM to do so. It's similar to how a carburetor works, but much more accurate and efficient.

omy, and increase power. A new, no-compromises small-block was needed.

Enter the LS1.

One look tells you the new small-block is nothing like the old one. In fact, apart from a few miscellaneous nuts and bolts, the only thing the Gen-III engine shares with either of the previous generations is the 4.40-inch center-to-center bore spacing, and even that was only retained because it happens to work well. Everything else about the engine, right down to the firing order is different (the old engine's hit on 1-8-4-3-6-5-7-2, while the Gen-IIIs light up to 1-8-7-2-6-5-4-3).

The upside of this clean-sheet design is that the LS1—and its iron-block brethren, the 1999 Vortec 4800, 5300, and 6000 V-8s—does not suffer many of the problems enthusiasts and professional engine builders have tried to outwit through the years.

For instance, the rocker arm geometry on the Gen-III engines lies in a single plane, eliminating complex compound angles. The camshaft is mounted higher in the block for added crankshaft and connecting rod clearance, and it features a larger core size, for added strength, yet a hollow center for reduced weight. The crankshaft has larger main bearing bores for reduced bearing stress, a longer stroke for increased torque, and redesigned journals with compression-hardened fillet areas instead of massive radii. The connecting rods are longer too, to increase piston dwell time at top dead center for more power, plus they feature cap screw-style bolts and a unique manufacturing process that forms the entire rod as one piece from powdered metal, then snaps off the cap in such a way that it requires no machining and prevents improper mating of caps to rods.

The block itself is almost a work of art. It's a deep-skirt design that enshrouds the crankshaft entirely within the block and holds it there with six main bearing cap bolts—four vertically and two (one from each side) horizontally to tie the caps firmly into the structure of the block for added rigidity. Up top, tall ribs connect the two cylinder banks in the lifter valley, while deck surfaces have four cylinder head bolt holes per cylinder—an arrangement that clears the way for an easy upgrade to four valves per cylinder on the engine.

The current two-valve heads, however, are marvels of performance. With extremely tall, narrow cathedral-style intake ports, the heads provide an incredibly efficient path for the fuel injectors to spray into the cylinders, past the 15-degree valves. Meanwhile, a composite (plastic) intake manifold sports exceptionally smooth airflow through its medium-length runners.

One of the nicest touches of the Gen-III engines is that there isn't a conventional gasket anywhere in the engine. Every gasket (except the cylinder head gaskets) is a rubber-like O-ring-style gasket sandwiched between aluminum surface plates that provide just the right crush without deformation for a perfect seal every time. Even more handy, however, is the fact that the gaskets can even be removed and reinstalled several times before they are no longer usable, which

The system that replaced TBI on production trucks is called central-port fuel injection (CFI), and it is similar to a TPI system.

CFI systems use a monoblade throttle body to control airflow into the engine. The intake manifold is a two-piece affair with a composite plenum and an aluminum base. A single fuel injector distributes fuel to a poppet valve in each intake port via flexible fuel lines.

should save hot-rodders a few bucks when swapping parts.

In production trim, the LS1 is a stout engine—345 horsepower. More impressive, however, are the power figures aftermarket tuners are getting out of the engines. Most notably, Callaway Cars of Old Lyme, Connecticut, features modified LS1 engines in its C12 grand touring sports car and develops 450 horsepower with only 5.7 liters of displacement. The engine idles like a stock LS1 but pulls like a big-block from the 1960s when the throttle is snapped.

Performance Potential

The Gen-III engines are already phenomenal performers and the aftermarket tuners show there's much more potential in the engine; however, the engine is so new and so few enthusiasts have them yet that aftermarket companies have not yet developed anywhere near the vast array of performance parts that exist for the previous two small-blocks.

Now that the engines are available in massquantities in GM's Chevy Silverado and GMC Sierra full-size truck lines, companies will undoubtedly begin spending more time developing upgrades for the engines.

Until that time, the engines that are modified require expensive custom-engineered components, which precludes most enthusiasts from doing more than adding the few bolt-on parts that are available, or contracting one of the tuners to build them a hopped-up LS1/'99 Vortec to go.

Vortec V-6 and 1996–1998 Vortec V-8 Engines

The original Vortec engines, in both V-6 and V-8 configurations (except for the Vortec 7400), are based on the late-style first-generation engine blocks, with roller camshafts and one-piece rear main seal crankshafts. (The Vortec 7400 is based on the Mark V big-block V-8 with similar modifications.) As such, and in keeping with Chevrolet's long-standing advertising slogan for its truck line, the Vortec engines are solid as a rock.

The Vortec V-8s, which were introduced in 1996 Chevrolet C/K full-size pickups, feature some of the best small-block Chevy cast-iron cylinder heads ever offered on production vehicles. Small, fast-burn combustion chambers are fed by tall LT1-style trapezoidal intake ports that are narrower at the top than the bottom to promote not only excellent flow but to maintain fuel atomization.

The Vortec's induction system has a great deal in common with the TPI V-8s. A large, round throttle body assembly feeds into a composite plenum that attaches to an aluminum base. Fuel is added by poppet valves mounted in each runner, just ahead of the cylinder head ports. A single fuel injector supplies fuel to the poppet valves via flexible fuel lines.

The engines are built on proven designs and develop excellent power. Unfortunately, parts are not widely available for modifications, beyond performance exhaust systems, better-flowing air filters, and some high-performance ignition components. Hard parts, such as camshafts and upgrades for the fuel

This 3.1-liter V-6 uses a very TPI-like system with an interesting cross-ram plenum.

delivery system, aren't widely available for the engines and aren't likely to be, given their short lifespan. Since they are based on the traditional small-block, the Vortecs could be given a hotter camshaft, such as from a TPI-style engine or an aftermarket piece, though that is likely to upset emissions controls. And because the Vortec engines feature OBD-II engine diagnostics, such well-intentioned upgrades will likely only hurt performance, unless the PCM can be reprogrammed to take advantage of them.

In the end, the Vortec engines, as excellent as they are, are best left in stock form. Enthusiasts will realize more significant improvements from less time spent on a more readily modifiable TPI-style engine.

Throttle Body Injection (TBI) Engines

Chevrolet's first attempts at electronic fuel injection were little more than glorified carburetors outfitted with fuel injectors suspended above their venturis. These so-called "wet flow" systems flowed both air and fuel through the intake manifold, which resembled a traditional intake manifold as would be used with a carburetor.

The systems first appeared in 1982 on Corvettes, Camaros, and Firebirds as a somewhat intriguing dual-TBI arrangement atop a cross-ram manifold. The system was given the dubious name "Cross-Fire Injection" and proved to be not quite ready for prime time. Chevy quickly got its act together, however, and by 1985 was using TBI on a broad variety of engines from lowly four-cylinders to full-size V-8-powered pickups. Those same basic designs would power Chevy trucks until the 1996 Vortec V-8s arrived. The TBI systems were inexpensive for Chevy to build and cost consumers virtually nothing in terms of service, requiring little in the way of maintenance and rarely needing repairs.

While the systems were far more fuel efficient than carburetors, they still suffer many of the same problems that make carburetors so inadequate. Their wet-flow design results in uneven mixture distribution to the various cylinders, as fuel falls out of suspension en route to the end cylinders, causing them to run leaner than the center cylinders. The fuel that does fall out of suspension forms small streams along the intake runner floors, and those streams eventually find their way into the cylinders, further complicating the fuel-mixture/cylinder balance problem.

The fact that TBI systems are so similar to carbureted systems does make

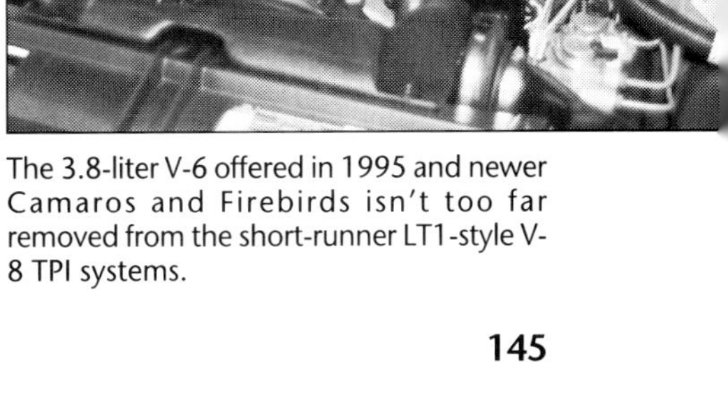

The 3.8-liter V-6 offered in 1995 and newer Camaros and Firebirds isn't too far removed from the short-runner LT1-style V-8 TPI systems.

Meanwhile, the initial Vortec V-6 truck engines were obviously patterned after the long-runner, torque-enhancing TPI plenum, runners, and intake manifold base.

The ZR-1 Corvette's LT5 was another direct descendent of the original, long-runner TPI intake system, but with some significant changes, such as two runners per intake. The system operated on a single runner until power demands warranted opening the second runners to increase airflow into the engine for more power.

them an excellent upgrade path for vintage vehicles, however. Holley Performance Products manufactures a production-like TBI system called Pro-Jection which is highly tunable and can be easily outfitted on carbureted vehicles. They can typically be installed in stealthy fashion, too, preserving the looks of vintage vehicles, while providing some of the many advantages of modern EFI engines, including quicker starting, better throttle response, better driveability, better fuel mileage, decreased exhaust emissions, and more.

Because of the popularity of truck customization, a number of performance parts have sprung up for the TBI truck engines, and while many of these same components would work in a TBI-equipped Camaro or other production passenger car, they are not generally emissions-legal for such applications. So, unless your vehicle is a TBI-equipped pickup (or van), odds are good that you would be more satisfied following a swap to a TPI-style induction system.

The Gen-III small-block features the latest iteration of the TPI induction system. The primary differences are a monoblade throttle body and a composite, medium-length runner intake manifold.

These are the major components of Holley's Pro-Jection throttle body injection EFI system (clockwise, from left): the ECM, a four-injector throttle-body unit and a disk to customize the fuel map for optimal performance.

CHAPTER ELEVEN

RETROFITTING AND CUSTOM INSTALLATION TIPS

BOLTING IN EFI ENGINES AND WIRING THEM UP

No sooner had Chevrolet officially unveiled the TPI engine in the 1985 Corvette, than car enthusiasts—street-rodders, truck customizers, owners of vintage muscle cars and hot 1950s classics—started using the engine to improve their cars. The Chevy TPI engines are naturals for such swaps, because they offer not only performance, but reliability and even an attractive, somewhat distinctive appearance too.

When it comes to retrofitting a Chevy TPI engine into your project vehicle, there are two basic steps to take: First, you need to get it physically bolted in place, and second, you have to make it run. As you'll soon learn, the former is most definitely the easier aspect of the job.

Bolting It In

Because the TPI, Gen-II, and pre-1999 Vortec engine series are based on the legendary small-block Chevy, the job of bolting them in place is fairly straightforward. In fact, if you happen to be swapping one of these EFI Chevys for a carbureted small-block, the new engine will literally bolt right in. It'll even mate with your existing transmission, since the bellhousing bolt patterns are the same.

But even totally custom swaps—such as dropping a TPI engine into a 1930s street rod—can generally be handled with ease, since you can often take advantage of motor mounts and various accessories that are widely available for use with traditional small-blocks. The first-generation TPI engines can even use a front motor plate, if necessary, for totally custom applications. And while the front end of the Gen II engines (LT1, LT4, and L99) is different enough that standard small-block motor plates won't work, the Gen-II engines have been around long enough (remember, the LT1

The original TPI engines—the L98 350 and LB9 305—are stylish, powerful, reliable engines, which makes them ideal engines to swap into older vehicles, as this 1957 Chevy's engine bay shows.

Big-block engines are always popular swap candidates, and when outfitted with an aftermarket TPI-style induction system like this 454's TPI Specialties Big Ram system, they're outstanding performers.

Here's another of Chevy's engineering exercises—a 510-inch big-block with a custom-fabricated TPI-style induction system based on a modified tunnel ram intake manifold. While the custom intake may be a bit complicated for most people's skills, aftermarket systems are available that make the swaps a breeze.

came out in 1992) that many specialty companies have developed kits and individual components that make installing them in custom applications far easier than hand-fabricating mounts and such from scratch.

The newest Chevy small-blocks, the LS1 and the 1999-and-newer Vortec V-8s based on it, are completely different animals, so swapping them is a little less cut-and-dried, but since the bellhousing bolt pattern is the same, at least the engine can be connected to the usual GM transmissions without the need for adapters. And while the Gen-III engine does feature different engine-mounting brackets, anyone competent with a welder should be able to either adapt the production mounts to their vehicle, or fabricate their own mounts.

Making It Run

Part of what makes Chevrolet's TPI engines—and all its EFI engines, for that matter—so reliable, powerful, and efficient is the electronics that control them. Unfortunately, those electronics also make the engines a good deal harder to swap into a car.

In the old days, before emissions controls became required, swapping an engine into a car simply required bolting it in, getting the ignition system some power, and running a fuel line to feed the engine. But EFI engines have (at the very least) an engine control module (ECM) or a later-style powertrain (or vehicle) control module (PCM or VCM) that not only dictates how the engine operates but also how the transmission works and even how the emissions controls function. In order for the control module to be able to do its job, it has to gather input from an array of sensors located on the engine and transmission and may control a number of servos and motors too. All these electronics and the spaghetti of wires that run between them and the control module make a swap much more confusing and require some modifications of some of the components you'll be using, such as adding oxygen sensor fittings to your headers.

Initially, few people understood the complex-looking electronics systems, and even fewer knew how to adapt the

Open and Closed Case

In the olden days, when TPI was new, enthusiasts didn't understand how the system worked, but it looked so cool people just had to have the systems on their hot rods and customs.

But because people didn't understand the system, many people installed the units without using oxygen sensors, thus the systems only operated in the "limp home" open-loop mode. Recalibrated PROM chips were used to adjust fuel curves, but the systems generally delivered poor performance and they weren't emissions legal.

Today, however, EFI systems are much better understood, and it's clear that the systems deliver far better power when they operate in the closed-loop mode, when they are able to adjust the air/fuel mixture based on the oxygen content of the exhaust gases. This ability makes the engine more powerful and cleaner running too.

Whether you're swapping an EFI engine to a non-EFI vehicle, or just an EFI induction system onto a non-EFI engine, there are wiring harnesses and electronic management systems that make it quick and easy to get the engine hooked up and running in closed-loop mode. In fact, it as easy now to make a system run in closed-loop mode the way it was designed to run than it is to make it operate the old way, in open-loop mode only.

Depending on the type of swap you are doing, it may not only be prudent to make sure your new EFI engine runs in closed-loop mode. It may be legally required, thanks to emissions laws policed by the EPA and state agencies, such as California's Air Resources Board.

The twin, cross-ram plenums and runners of this exotic-looking dual-throttle-body-equipped multipoint injection system tops an equally exciting engine—a 572-cubic inch, 720-horsepower all-aluminum big-block. It was shoe-horned into a 1993 Camaro by Chevrolet's Raceshop skunk works to show what 1960s muscle cars would be like with today's technology.

While the valve covers resemble those of a big-block—and so does its displacement—this custom installation pushes the envelope with its all-aluminum 427-cubic inch NASCAR-inspired "SB2" engine with a home-brewed TPI-style induction system. The engine was built by Lingenfelter Performance Engineering and thumped out 600 horsepower and 560 pound-feet of torque! It's installed in a 1995 Corvette.

Organ Donors

One of the easiest and most cost-effective ways to swap an EFI engine is to find a car that is equipped with the engine you want and swap everything over to your vehicle. Used Camaros and Firebirds—especially ones that have been involved in rear-end or side collisions—are often available at quite reasonable prices, and you can get not only the engine, but the PCM, wiring harnesses, transmissions, and other parts that make a swap easier.

systems to custom installations where the engines were likely to feature various performance parts, such as hotter camshafts or better flowing heads. Even just the presence of headers could be enough to throw those initial control systems out of whack. So people figured out how to simply make the engine run in what GM calls the "limp home" mode, or more technically the "open loop" mode, referring to whether or not the computer was using input from the oxygen sensor plumbed into the exhaust system to determine whether the air/fuel mixture was too rich or lean. A closed-loop system uses the oxygen sensor, while an open-loop system does not and automatically defaults to a richer fuel mixture to prevent detonation that could damage the engine. Eventually, people realized that running a TPI-style engine in open-loop mode didn't make much sense, since the engine wasn't able to vary its performance to suit different driving conditions, plus the engine operated inefficiently and was hard to modify for improved performance.

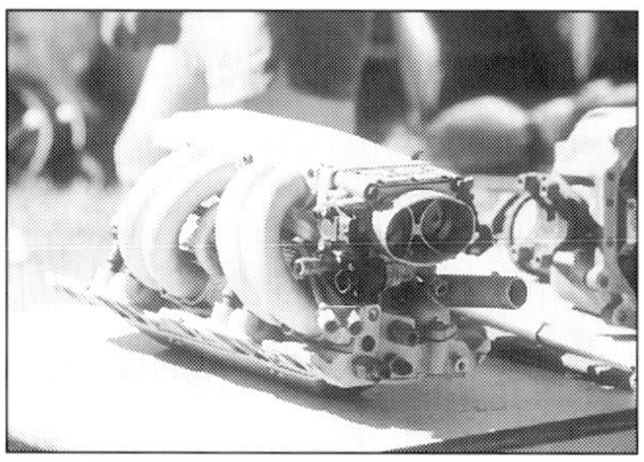

Of course, you can also go the route of simply swapping your engine's existing induction system for a TPI system. This is a stock, long-runner TPI system from either an L98/B2L 350 or an LB9 305.

Making matters even worse for swappers is the fact that the control modules are often so tightly integrated into the rest of the vehicle that it can be nearly impossible to retrofit the electronics to your project car.

Fortunately, several aftermarket companies offer electronic control systems designed specifically to ease custom engine installations. While that alone is cause for celebration, even more beneficial is the fact that the aftermarket control systems can often be tuned with a personal computer to alter the fuel and ignition system curves to compensate for any advanced engine components you may have installed or any special driving conditions. In fact, some of the systems can be hooked up to a laptop computer and adjusted as you drive, while the latest systems feature dedicated controls that eliminate the need for a personal computer at all. Reprogramming your fuel or ignition curve is like changing a station on your stereo: just push a button.

Beside the aftermarket control system's actual control module, the kits typically include a wiring harness with the proper leads for all the necessary sensors, motors, and other devices to which it must connect.

Little Things Mean a Lot

When swapping in an EFI Chevy engine, there are a number of little things that can make a big difference in not only how easy your swap goes, but how well your engines runs (or even whether it will run at all!) once it's installed.

There are a number of ways to go about swapping in TPI power. An obvious one is to simply pick up a complete TPI engine. These were found at the "Corvettes At Carlisle" swap meet in Carlisle, Pennsylvania.

Fuel for Thought

One of the most overlooked aspects of a TPI engine swap is the fuel system. Again, reflecting on yesteryear's simplicity, carbureted fuel systems were simple: A fuel tank stored the gas, a steel line carried it to the mechanical fuel pump, which then sent the fuel to the carburetor. EFI systems, however, use electric fuel pumps of much higher capacity than a mechanical unit. TPI fuel pumps, in particular, flow a lot of fuel to the engine at very high pressure to ensure a good spray pattern from the injector nozzle tip. Because the fuel is under high pressure, it would rapidly heat up were it not for the fact that GM engineers incorporated a fuel return line, so fuel can continuously flow from the tank to the injectors and back to the tank.

Not too Cool, Not too Hot

Swappers also need to be aware of at least two specific issues with regard to cooling systems: thermostat temperatures and water pump rotation direction. One engine parameter that the control module monitors is coolant temperature, which indicates to the engine whether the engine is warmed up or not, and thus whether it is ready to run in closed-loop mode or not. Muscle car engines from the 1960s ran best when they stayed cool: It minimized detonation and kept the intake air charge from getting too hot. So, for those engines, it was common to run a thermostat that opened at 160 degrees Fahrenheit. But cooler engine temperatures produce more harmful exhaust emissions than do hotter temperatures, so most Chevy EFI engines run hotter 190- to 195-degree

Aftermarket EFI Systems

Chevy's production electronic fuel-injection systems are good—very good. But they aren't perfect for every application, nor are they even available for every application.

Fortunately, the aftermarket performance parts industry has stepped in to provide solutions that many enthusiasts might need or find appealing.

As with Chevrolet's stock systems, two types of aftermarket systems are available: throttle-body injection or multipoint. Each is available for both small- and big-block applications.

Aftermarket throttle-body injection systems can often be used with traditional carbureted intake manifolds, though some performance TBI intakes are available to take better advantage of the performance potential of the more precise fuel metering. Multipoint systems require their own specific manifold designs.

Holley Performance Products introduced its Pro-Jection throttle-body injection-style systems several years ago as easy, effective, and affordable methods of upgrading carbureted vehicles to EFI. The systems come complete with everything needed to convert to EFI, including the throttle body assembly, ECM, wiring harness, and even a fuel pump and easy-to-follow instructions.

Edelbrock manufactures a multipoint system that is available to fit small- and big-block engines alike. The systems include an intake manifold outfitted with fuel distribution logs and individual injectors plumbed into the ends of the runners. As with the Holley systems, the computer, wiring harness, fuel pump, and other required items are included to make the swap as easy as possible.

TPI Specialties offers a multipoint injection system for Mark IV and Gen-V big-block Chevrolet engines, which provides the many benefits of a dry-flow EFI system for performance big-block engines like the 396, 427, and 454.

ACCEL also specializes in multipoint systems. It offers complete TPI-style systems with oversized runners and ported intakes and plenums to match. It also has a special John Lingenfelter–designed system called the SuperRam, which features a large plenum, medium-length runners, and a matching base. In addition, it offers two big-block systems depending on your needs. ACCEL also offers a wide variety of EFI sensors and other components that not only work with its fuel-injection systems but are excellent replacements and upgrades for stock equipment as well.

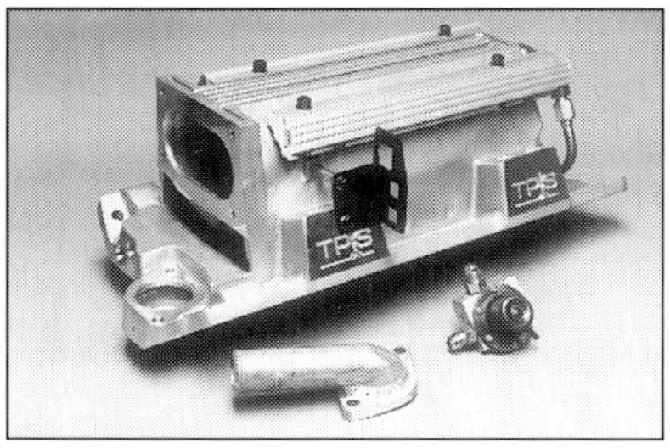

Another option for small-block engines is to go with a short-runner intake like TPI Specialties' MiniRam intake, which is a virtual clone of the production LT1/LT4 intake. Be aware that differences between the three generations of small-block Chevy V-8s dictates specific intake designs for each; intakes are not interchangeable between the different engine designs.

ACCEL's TPI-style Stealth intake system is designed to increase low- and mid-range torque on big-block engines and is available with everything needed for an easy installation.

Another aftermarket solution that makes for an easy upgrade to EFI is Holley's Pro-Jection Throttle-Body Injection system, which comes with everything needed to upgrade from a carburetor to EFI, including the ECM, wiring harness, high-pressure fuel pump, and more. The TBI unit can be mounted to an existing intake manifold (with Holley square-bore mounting flange), or Holley has an intake specially designed to make the most of the Pro-Jection system.

thermostats. Of course, that extra heat does cause the engine to be closer to a point where detonation sets in, and it also warms the intake air more (on non-Gen-II engines), which causes the air to be less dense, so there is less oxygen for fuel to adhere to and thus the engine makes less power. But changing to a 160-degree thermostat can cause the engine's control module to think the engine isn't fully warmed up yet, and thus isn't ready to run in closed-loop mode (except input from the oxygen sensor to tailor the fuel curve). To avoid that, some companies offer 170-degree thermostats that allow the ECM to detect a fully warmed-up engine so it can operate in the closed-loop mode.

Additionally, when Chevy switched to a serpentine belt to turn the engine accessories (power steering pump, air conditioning, water pump, and so on), the company had to reverse the rotational direction of the water pump. So while V-belt engines turn the water pump in the same direction as the crankshaft (clockwise), serpentine belt engines turn the pump counterclockwise. The reason this is important is that the vanes on the water pump's impeller are directional; if you turn a pump designed for counterclockwise rotation clockwise, it will hardly pump any coolant through the engine, and the engine will always run hot—and probably overheat.

Heads, You Win . . .

While there's much more information about cylinder heads in chapter 3, there are a few quick things worth pointing out for engine swappers. Presumably for assembly purposes, GM changed the angle of the center intake manifold bolts and their corresponding tapped mounting holes in the cylinder heads in the late-1980s. The difference in angles is readily apparent, especially if you simply compare the angle of the center bolt

TPI Engines without the TPI

Late-model production engines, such as the LT1 and LT4, are great engines even aside from their induction systems, which makes them logical candidates for someone looking for an easy engine upgrade.

General Motors' Service Parts Operations sells complete crate engines as well as just the necessary parts to swap an LT1 or LT4 long-block assembly to an older vehicle with a carbureted induction system and a traditional rear-mounted distributor.

Because the Gen-II engines are so similar to the first-generation small-block V-8s, the swap is a straightforward bolt-in, requiring only a different accessory drive system (because of the Gen-II's different water pump design) and a flywheel or flexplate designed for use with the late-model one-piece rear main seal crankshaft.

If you're opting to swap a factory system, one of the best ways to do it is to get a complete parts car. This 1984 Corvette would not only get someone the cross-fire injection system but the wiring harness, ECM, and numerous other components that can be a real hassle to round up individually later on.

holes (in the heads) to those at the ends of the intake mating surface. This is important to know about because it can lead to a headache or two when trying to mount your favorite intake manifold. Most aftermarket intakes now feature center bolt holes that will work with either the early- or late-style heads, and it is possible to modify manifolds to work with either. But why create more work for yourself?

It's also not commonly known that cylinder heads for 305-cubic inch engines are different than those for the 350 engines. The specific difference that can cause problems has to do with the spacing of the valves; because the 305 has a smaller diameter bore than the 350, the valves must be located closer to the bore centerline to avoid possible contact with the block's deck surface or the cylinder walls. So, 350 heads can't be used on a 305. And while 305 heads could be used on a 350, why would you want to? The valves in a 305 head crowd and shroud each other, and you can't easily run oversize valves in a 305 head, which limits flow through the head even more.

Cranks for the Memories

Crankshafts feature even more differences than the cylinder heads. One thing to remember is that cranks with old-style two-piece rear main seals are internally balanced (except for the 400-inch production engine and custom stroker engines, like 377, 383, and 388-inch engines that use the 400 crank). Cranks for one-piece seals are externally balanced, thus they require different harmonic dampers and flywheels or flex plates than you would use for an internally balanced traditional first-generation 350, for example. Another issue with balance has to do with the 305 versus 350 cranks. Both feature a 3.48-inch stroke, but the 305 crank is lighter than the 350 crank, because the 305 piston and connecting rod assemblies are lighter and thus require less counterweight material to balance. It is possible to run a 305 crank in a 350, but you may run into balancing problems, especially if you've changed to forged pistons or heavier-duty connecting rods. The 305 crank can be balanced with heavier piston/rod assemblies by adding heavy metal, but this is expensive and many machine shops prefer not to do it. Your best bet, then, is to locate a good 350 crank. Or better yet, buy a true high-performance crank from an aftermarket crank manufacturer for even better performance.

Mean and Clean

Unless you've been lost in space for the last few decades, you can't help but be aware of the federal Environmental Protection Agency (EPA) and California's

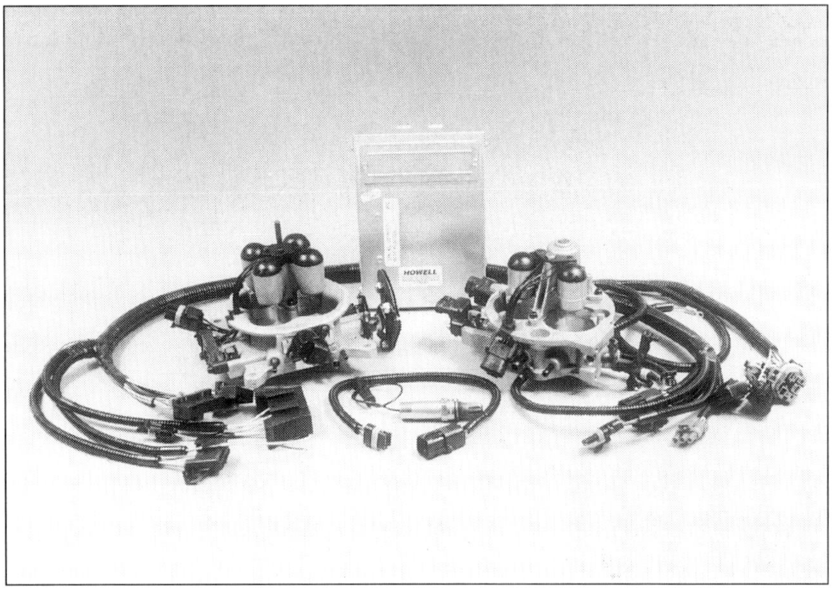

Like Holley, Howell Engine Developments has a TBI system swap kit, which features a GM-style TBI unit (with either two or four injectors), the wiring harness, ECM, and other miscellaneous parts.

Air Resources Board (CARB) and their disapproval of unhealthy engine exhaust emissions. Contrary to popular belief, however, the EPA and CARB aren't evil ogres intent on robbing automotive enthusiasts of their fun. They just ask that we stay "clean and green" when doing it.

The EPA and CARB each forbid any changes that modify, eliminate, or otherwise render inoperable any emissions control equipment. They do, however, grant aftermarket companies "executive order exemptions" for equipment the manufacturers can prove don't violate the agencies' prime directive.

It's also important to understand that the EPA and CARB require a vehicle to meet emissions output levels for that year, make, and model vehicle—not necessarily for the engine that is installed in it. So, if you swap an EFI engine into a 1967 Camaro, you still should only have to meet 1967 emissions standards. If you happen to want to swap a 1991 TPI engine into a 1998 Chevy pickup, for some reason, you're going to have your work cut out for you getting the old engine to meet the much less lenient 1998 levels of acceptable emissions. Also, just because you can swap the engine (with a like-design engine), that doesn't mean you can swap everything else connected to that engine. For instance, 1982 to 1999 Camaros have not been available from Chevy with a true dual-exhaust system that uses twin pipes from the engine all the way back to the tailpipes; instead, they use a single, large intermediate pipe from the catalytic converter to the muffler. So, while it's technically possible, it's highly illegal to cobble together a dual-exhaust system for these cars or any other that wasn't available with one from the factory.

In most cases, late-model engines are swapped into early-model cars, and—assuming all the emissions controls devices are retained in proper functioning order—the result is far cleaner emissions for the vehicle at its next tailpipe sniffer test.

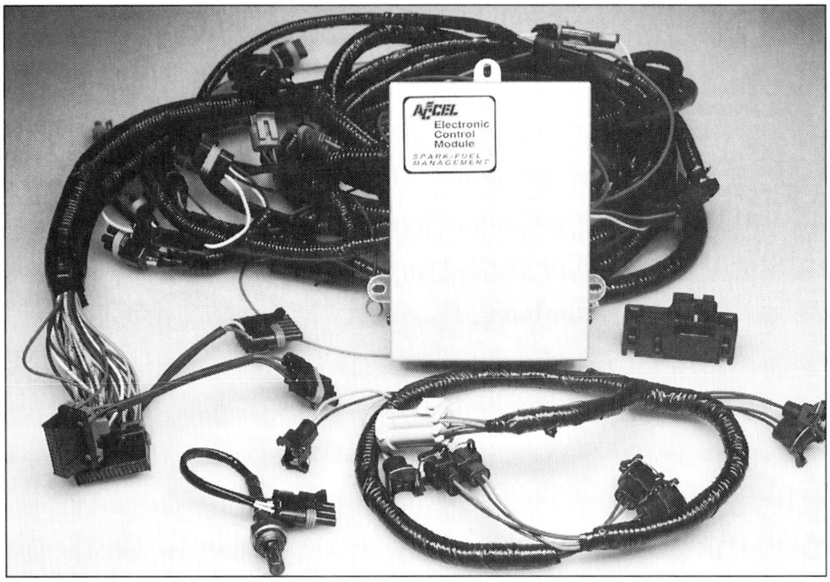

If you happen to be using a factory or aftermarket EFI system but need a computer and other components to make it work, ACCEL has a full line of "service" parts, including everything from individual injectors to easily programmable high-performance ECMs.

APPENDIX A: PARTS AND SERVICES SOURCES

ACCEL
8700 Brookpark Road
Cleveland, OH 44129
216-398-8300
www.mrgasket.com
Ignition, induction, and fuel-injection systems and components

Airflow Research
10490 Ilex
Pacoima, CA 91331
818-890-0616
Cylinder heads

Autotronic Controls Corporation/MSD Ignition
1490 Henry Brennan Drive
El Paso, TX 79936
915-857-5200
Ignition systems and components

BBK Performance Parts
1611 Railroad Street
Corona, CA 91720
909-735-2400
Induction and exhaust components

Borla Performance
5901 Edison Drive
Oxnard, CA 93033
805-986-8600
www.borla.com
Exhaust systems

Callaway Cars
3 High Street
Old Lyme, CT 06371
860-434-9002
www.callawaycars.com
Performance engine building services and components

Centerforce Clutch
2266 Crosswind Drive
Prescott, AZ 86301
520-771-8422
Clutches

Competition Cams
3406 Democrat Road
Memphis, TN 38118
901-754-2400
www.camhelp.com
Camshafts, valvetrain components

Crane Cams
530 Fentress Boulevard
Daytona Beach, FL 32114
904-258-6174
Ignition system and valvetrain components

Doug Rippie Motorsports
14070 23rd Avenue North
Plymouth, MN 55447
612-559-7605
Performance engine building services and components

DynoMax Performance Exhaust
(Contact your local performance parts dealer)
Exhaust systems and components

Earl's Performance Products
189 West Victoria Street
Long Beach, CA 90805
310-609-1602
Braided hoses and lines

Eastwood Company
580 Lancaster Avenue
Box 296
Malvern, PA 19355
800-345-1178
Tools, restoration supplies

Edelbrock Corporation
Torrance, CA 90503
310-782-2900
Cylinder heads, intake manifolds/runners, throttle bodies, exhaust systems, suspension components

Fel-Pro Performance Products
One Equion Drive
Ashland, MS 38603
601-224-8972
Gaskets

Flowmaster Exhaust
2975 Dutton Avenue
Unit #3
Santa Rosa, CA 95047
707-544-4761
Exhaust systems

GM Performance Parts
(See your local General Motors dealer's parts department)

Griffin Radiator
Piedmont, SC 29673
800-RACE RAD
Radiators

Hays Clutches
8700 Brookpark Road
Cleveland, OH 44129
216-398-8300
www.mrgasket.com
Clutches

Hi-Performance Coatings (HPC)
Salt Lake City, UT
800-456-4721
www.hpcoatings.com
Thermal and corrosion protection coatings

Holley Performance Parts
P.O. Box 10360
Bowling Green, KY 42102-7360
502-781-9741
Cylinder heads, induction and ignition systems, and components

Hooker Industries
1024 W. Brooks Street
Ontario, CA 91762
909-983-5871
Exhaust systems and components

HP Motorsports
5055 S. 36th Street
Omaha, NE 68107
402-731-7301
www.hpmotorsport.com
Engine components

HyperTech
1910 Thomas Road
Memphis, TN 38134
901-382-8888
Induction, injection, and computer components

Jacobs Electronics
500 N. Baird Street
Midland, TX 79701
800-627-8800
www.jacobselectronics.com
Ignition systems and components

JBA Headers
7149 Mission Gorge Road
San Diego, CA 92120
619-229-7797
www.jbaheaders.com
Exhaust headers

Jet Hot Coatings
610-277-5646
Thermal and corrosion protection coatings

K&N Engineering
561 Iowa Avenue
Riverside, CA 92507
800-858-3333
Air filters

LG Motorsports
4314 Action Street
Garland, TX 75042
972-272-7753
Engine components

Lingenfelter Performance Engineering
1557 Winchester Road
Decatur, IN 46733
219-724-2552
www.lingenfelter.com
Performance engine building services and components

March Performance Pulleys
5820 Hix Road
Westland, MI 48185
313-729-9070
www.marchperf.com
Engine pulleys

Moroso
80 Carter Street
P.O. Box 1470
Guilford, CT 06437-0570
203-453-6571
Numerous components

Mr. Gasket
8700 Brookpark Road
Cleveland, OH 44129
216-398-8300
www.mrgasket.com
Accessories

Nitrous Oxide Systems (NOS)
5930 Lakeshore Drive
Cypress, CA 90630
714-821-0580
Nitrous oxide systems

Nitrous Works
1450 McDonald Road
Dahlonega, GA 30533
706-864-7009
Nitrous oxide systems

Paxton Products
1260 Calle Suarte Street
Camarillo, CA 93012
805-987-5555
Superchargers

Random Technology
1313 Temple Johnson Road
Loganville, GA 30052
770-978-0264
High-performance induction and exhaust system components

Redline Synthetic Oil Corporation
3450 Pacheco Boulevard
Martinez, CA 94553
510-228-7576
Synthetic lubricants

Rinda Technologies (Diacom Plus)
4563 N. Elston Avenue
Chicago, IL 60630
773-736-6633
Computer (ECM/PCM) interface system

RK Sport
214 Via El Centro
Oceanside, CA 92054
760-433-1663
www.rksport.com
Engine components and accessories

Russell Performance Products
225 Fentress Boulevard
Daytona Beach, FL 32114
800-394-1120
Braided hoses and lines

SLP Engineering
1501 Industrial Way North
Toms River, NJ 08755
908-349-2109
www.slpeng.com
Engine, suspension, brakes, body, interior components, and accessories

Summit Racing Equipment
P.O. Box 909
Akron, OH 44309
800-230-3030
Numerous performance and racing components

TPI Specialties
4255 County Road 10 East
Chaska, MN 55318
612-448-6021
www.tpis.com
Engine assemblies and components

Trick Flow Specialties
1248 Southeast Avenue
Tallmadge, OH 44278
330-630-1555
www.trickflow.com
Cylinder heads

Vibratech (Fluidampr)
537 E. Delavan Avenue
Buffalo, NY 14211
716-895-8000
Harmonic dampers

Vortech Engineering
5351 Bonsai Avenue
Moorpark, CA 93021
805-529-9330
Superchargers

World Products
29 Trade Zone Drive
Ronkonkoma, NY 11779
810-244-9822
www.goracing.com/world/
Cylinder heads

APPENDIX B: DIMENSIONS AND SPECIFICATIONS

Cylinder Blocks

Displacement (ci)	Model Years	Bore (in)	Main Journals (in)	Main Cap Bolts
262	1975–76	3.671	2.45	2
265	1955–56	3.75	2.30	2
267	1979–81	3.50	2.45	2
283	1957–67	3.825	2.30	2
302	1967	4.00	2.30	2
302	1968–69	4.00	2.45	4
305	1976–97	3.736	2.45	2
307	1968–73	3.825	2.45	2
327	1962–67	4.00	2.30	2
327	1968–69	4.00	2.45	2
350-I	1967–86	4.00	2.45	2 & 4
350-I*	1986–97	4.00	2.45	2 & 4
350-II	1992–97	4.00	2.45	2 & 4
350-V	1992–97	4.00	2.45	2 & 4
400	1970–72	4.125	2.65	4
400	1973–80	4.125	2.65	2

Key: I = Gen-I; II = Gen-II; III = Gen-III; V = Vortec; * = One-piece rear main seal

Crankshaft and Main Bearings

Stroke lengths

Displacement (ci)	Model Year	Stroke (in)	Main Journal Diameter (in)	Rod Journal Diameter (in)	Rear Seal Type
262	1975–76	3.10	2.45	2.10	2 pc.
265	1655–56	3.00	2.30	2.00	2 pc.
267	1979–81	3.48	2.45	2.10	2 pc.
283	1957–67	3.00	2.30	2.00	2 pc.
302	1967	3.00	2.30	2.00	2 pc.
302	1968–69	3.00	2.45	2.10	2 pc.
305	1976–87	3.48	2.45	2.10	2 pc.
305	1988–97	3.48	2.45	2.10	1 pc.
307	1968–73	3.25	2.45	2.10	2 pc.
327	1962–67	3.25	2.30	2.00	2 pc.
327	1968–69	3.25	2.45	2.10	2 pc.
350-I	1967–86	3.48	2.45	2.10	2 pc.
350-I, V	1987–97	3.48	2.45	2.10	1 pc.
350-II	1992–97	3.48	2.45	2.10	1 pc.
400	1970–80	3.75	2.65	2.10	2 pc.

Main Bearing Clearances

1, 2, 3, 4, 5	0.002–0.003-inch (min. preferred)
End play	0.005-0.007-inch
Run-out (bend)	≤ 0.003-inch

Pistons and Rings

Piston-to-bore	0.005–0.0055-inch*
Piston-to-deck	
w/ 0.018-inch gasket	0.015–0.020-inch below deck
w/thicker gasket	0.035-inch (min.)
Piston-to-valve	0.020-inch (min., <8,000 rpm)
at 0 valve lash; 0.100-inch preferred	
Ring end clearance	
Top	0.022-inch
Second	0.016-inch
Oil	0.016-inch
Wrist Pins	
In piston	0.0004–0.0008-inch
In rod (floating)	0.0005–0.0007-inch
End play (floating)	0.0–0.005-inch

*Forged pistons; as measured at centerline of wrist pin hole, perpendicular to the pin.

Connecting Rods

Bearing clearance	0.002–0.0025-inch
Side clearance	0.010–0.020-inch

Cylinder Head/Intake Manifold Milling Guide

Cylinder Head (in)	Intake Side (in)	Intake Bottom (in)
0.005	0.006	0.009
0.010	0.012	0.017
0.015	0.018	0.025
0.020	0.025	0.034
0.025	0.031	0.043
0.030	0.037	0.052

INDEX